Reviews and critical articles covering the entire field of normal anatomy (cytology, histology, cyto- and histochemistry, electron microscopy, macroscopy, experimentaı morphology and embryology and comparative anatomy) are published in Advances in Anatomy, Embryology and Cell Biology. Papers dealing with anthropology and clinical morphology that aim to encourage co-operation between anatomy and related disciplines will also be accepted.
Papers are normally commissioned. Original papers and communications may be submitted and will be considered for publication provided they meet the requirements of a review article and thus fit into the scope of "Advances". English language is preferred, but in exceptional cases French or German papers will be accepted.
It is a fundamental condition that submitted manuscripts have not been and will not simultaneously be submitted or published elsewhere. With the acceptance of a manuscript for publication, the publisher acquires full and exclusive copyright for all languages and countries.
Twenty-five copies of each paper are supplied free of charge.

Manuscripts should be addressed to

Prof. Dr. F. **BECK,** Department of Anatomy, University of Leicester, 6 University Road, GB-Leicester LE1 7RH

Prof. W. **HILD,** Department of Anatomy, Medical Branch, The University of Texas, Galveston, Texas 77550/USA

Prof. Dr. J. van **LIMBORGH,** Universiteit van Amsterdam, Anatomisch-Embryologisch Laboratorium, Mauritskade 61, Amsterdam-O/Holland

Prof. Dr. R. **ORTMANN,** Anatomisches Institut der Universität, Lindenburg, D-5000 Köln-Lindenthal

Prof. J.E. **PAULY**, Department of Anatomy, University of Arkansas for Medical Sciences, Little Rock, Arkansas 72205/USA

Prof. Dr. T.H. **SCHIEBLER,** Anatomisches Institut der Universität, Koellikerstraße 6, D-8700 Würzburg

Advances in Anatomy
Embryology and Cell Biology

Vol. 75

Editors
F. Beck, Leicester W. Hild, Galveston
J. van Limborgh, Amsterdam R. Ortmann, Köln
J.E. Pauly, Little Rock T.H. Schiebler, Würzburg

Volker Grouls Burkhard Helpap

The Development of the Red Pulp
in the Spleen

With 37 Figures

Springer-Verlag
Berlin Heidelberg New York 1982

Dr. Volker Grouls, PD
Prof. Dr. Burkhard Helpap
Institute of Pathology
University of Bonn
Postfach 21 20
5300 Bonn 1
F.R.G.

Supported by Deutsche Forschungsgemeinschaft
Bonn-Bad Godesberg, He 537/5

ISBN-13:978-3-540-11408-6   e-ISBN-13:978-3-642-68514-9
DOI: 10.1007/978-3-642-68514-9

Library of Congress Cataloging in Publication Data
Grouls, Volker, 1943 – The development of the red pulp in the spleen.
(Advances in anatomy, embryology, and cell biology; v. 75) Bibliography:
p. Includes index. 1. Spleen-Cytology. 2. Spleen-Growth. 3. Develop-
mental cytology. I. Helpap, Burkhard. II. Title. III. Title: Red pulp in
the spleen. IV. Series. [DNLM: 1. Spleen-Growth and development.
W1 AD433K v. 75/WH 600 G882d]
QL801.E67 vol 75 [QL868] [599.03'3] 82-3350
ISBN-13:978-3-540-11408-6 (U.S.)   AACR2

Composition: Schreibsatz Service Weihrauch, Würzburg

2121/3321-543210

Contents

1 Introduction

In many aspects hematopoiesis in newborn rodents, especially in rats, resembles hematopoiesis in the human fetus in the 6th–7th month of gestation. In man the transition from the stage of liver to bone marrow erythropoiesis takes place at this time (Bessis, 1973). In rodents, however, the liver is almost the only place where hematopoiesis occurs until birth. Thereafter it is replaced to a growing extent by the bone marrow, which so far consists mainly of immature mesenchymal cells (Maximow, 1910; Cuda, 1970). Thus hematopoietic precursor cells appear in the sternum only around 30 h after birth. Just as in premature human infants, a macrocytic anemia can be demonstrated in normal neonatal rats (Lucarelli et al., 1964, 1968).

Beside liver (fetal) and bone marrow, the spleen is involved in hematopoiesis. In rodents like rats and mice, splenic hematopoiesis persists more or less markedly until adulthood; in man, however, it ceases after birth and reappears only under certain pathological conditions (Fischer et al., 1970; Hennekeuser et al., 1967; Fresen, 1960). In mice a considerable extramedullary hematopoiesis in the spleen is normal; under pathological conditions (e.g., hemorrhage) it may account for 50% of the necessary increase in erythropoiesis (Boggs et al., 1969; Bozzini et al., 1970). On the contrary, extramedullary hematopoiesis in rats is continuously reduced and stops almost completely in adult animals (Hardy, 1967). Thus the different sites of hematopoiesis in rats exhibit the following age-dependent activities:

1. Bone marrow: rapid activation postnatally, main place of hematopoiesis for the rest of life
2. Liver: fast decreasing activity after birth, stop within 7–10 days
3. Spleen: gradual decrease in activity after weeks to months

A number of morphological and cell kinetic studies on the postnatal development of the hematopoiesis in the bone marrow of rats have been performed concerning in particular the stem cell problem (Harris and Burke, 1957; Fliedner et al., 1968a; Haas et al., 1967; Renricca et al., 1976). Another investigation deals with the possible kinds of control mechanisms of the short-term postnatal liver hematopoiesis (Haas et al., 1970).

By contrast, little information exists about the postnatal development of the extramedullary hematopoiesis in the rat spleen (Andrew, 1946; Kindred, 1940, 1942; Lord, 1965a). These investigations were carried out on paraffin sections or autoradiographs of spleen smears. However, while in paraffin sections cell classification is very difficult, in paraffin autoradiographs it is hardly possible at all. Single cells seen in smears do not allow a reliable correlation to their origin from the anatomical structures of the spleen. Furthermore, the detached cells are not necessarily representative of the red pulp, because cell populations more firmly adherent to the supporting connective tissue may not readily be released (Metcalf and Moore, 1971; Block, 1976).

For these reasons the present study seemed to be appropriate. It comprises a cell-analytical and cell-kinetical investigation of the physiological hematopoiesis and related stromal elements in the spleen of untreated rats, in correlation with age and growth development, carried out by means of semithin sections and semithin-section autoradiographs.

The concept was further supported by the fact that — considering its temporal activity — the extramedullary hematopoiesis in the spleen obviously stands in between the fast decreasing liver hematopoiesis and the postnatally developing bone marrow hematopoiesis. Thus it seemed to be possible that the hematopoiesis in the spleen might be subjected to a control mechanism and proliferation type different from that in liver and bone marrow.

Finally, the results obtained should be compared with the observations on spleen, liver, and bone marrow hematopoiesis so far described in the literature.

2 Autoradiography with ^{3}H-Thymidine (^{3}H-TdR)

The pyrimidine nucleoside thymidine is a specific precursor of DNA. Injected during the S-phase of the cell cycle, it is incorporated into the cell nucleus as an almost exclusively stable, not exchangable compound. By means of radioactively labeled thymidine and autoradiographic techniques, newly formed DNA can be demonstrated.

The ^{3}H-TdR incorporated into the nucleus emits low-energy β-particles which act on a sensitized photographic emulsion overlying the histologic section. By the reduction of positively charged silver atoms a latent image is formed, which becomes visible as a punctual aggregation of black metallic silver grains after the usual photographic development and fixation. Leftover tritiated thymidine that is not incorporated into the DNA within the so-called availability time, is rapidly metabolized into products that cannot be used for DNA synthesis. These are either excreted over body fluids or removed by tissue fixatives (Baserga and Wiebel, 1969). The "blackening" of the photographic emulsion, i.e., the number of recognizable silver grains, is directly proportional to the radiation intensity of the incorporated amount of radioactive thymidine and dependent also on the "exposure time."

Because of the above-mentioned properties, ^{3}H-TdR is generally considered to be an ideal substance for the investigation of the proliferation of cells and analysis of the cellular generation cycle and its subphases (Baserga and Wiebel, 1969; Maurer and Schultze, 1968).

In a homogenous cell population the percentage of labeled cells, opposed to the fraction of nonlabeled cells, is called "labeling index" or "^{3}H-index." Under steady state growth conditions it is dependent on the ratio of the duration of the DNA synthesis (t_s) to the duration of the entire cell cycle (t_c). Taking different variables into consideration, such as growth fraction, diurnal fluctuation, and growth conditions, the labeling index is — like the mitotic index — regarded as a measure for the proliferative activity of a tissue. Because of the longer S-phase, the labeling index is about tenfold the mitotic index and more accurate.

The generation time of the cells can be estimated from the counted labeling index and the known t_s determined in a different way. However, according to the growth pattern of the cell populations (steady state growth, exponential growth), certain reservations have to be considered, due to frequency distributions of the cells in the cell cycle (Maurer and Schultze, 1968). Consequently, different mathematical models of the hematopoiesis have been constructed, which either assume a steady state equilibrium (rectangular phase distribution in the different cell compartments) or an exponential growth pattern (logarithmic phase distribution; Blackett, 1971; Hanna and Tarbutt, 1971). However, the assumption of a steady state equilibrium of the perinatal hematopoiesis possibly does not mirror the true proliferative dynamics. Alternatively, it may be expected that the hematopoiesis in newborn rats actually shows an exponential growth pattern because of the rapidly expanding blood volume and an increased demand for a cellular defense against infectious agents after birth. However, it may well be that steady state conditions are already reached some weeks after birth (Haas et al., 1969a).

Similar to the epithelium of the intestinal crypts, in hematopoiesis a cell system with an exponential growth pattern is concerned. But due to the simultaneity of proliferation and maturation, a constant loss of cells from the proliferative compartment occurs. After completed maturation the cells leave the proliferative cell pool at a constant rate, corresponding to the rate of production, and transform into nondividing cells. If exactly as many cells per time unit leave the proliferation pool as are produced by mitosis, the number of cells and frequency distribution remain constant. This is the special case of exponential growth under steady state conditions (Maurer and Schultze, 1968).

Under exponential growth conditions, the mitotic index is smaller by the factor log n 2 than under steady state conditions (Maurer and Schultze, 1968; Hanna and Tarbutt, 1971), and is then given by the equation $MI = 0.693 \times t_m/t_c$.

The duration of the S-phase may be computed for exponentially growing cell populations from the equation $LI = \lambda \times t_s/t_c$ (Tarbutt and Blackett, 1968; Hanna and Tarbutt, 1971). λ is a numerical factor, whose magnitude depends on the relative proportion of cells at different stages in the cell cycle and is usually in the range of 0.7−1. However, the factor λ may be neglected because of the following reason: It has been shown diagrammatically that in the case of generation times shorter than 1 day, the S-phase is equally positioned approximately in the middle of the cell cycle for both exponential and steady state growth. Therefore, the labeling index is nearly identical under both sets of growth conditions (Maurer and Schultze, 1968). A critical analysis of phase distribution and morphological cell compartmentalization is given by Dörmer (1973).

A single injection of ^{3}H-TdR (pulse labeling) followed by a radioactive experiment of 1 h, allows an examination and estimation of the normal, local cell proliferation in different areas of the spleen. If longer time intervals after pulse labeling with ^{3}H-TdR elapse, an increase in the labeling indices is first observed, according to the divisional activity of the cell population. With each division the radioactivity is passed on to two daughter cells, and then the density of the silver grains is only half as high.

In most hematopoietic cell systems proliferation and maturation take place simultaneously, making the cells pass morphological compartment boundaries. After a period of time corresponding to the length of the premitotic (G_2) and mitotic phase, a gradual increase in the percentage of labeled cells − due to the division of precursor cells − can be expected (Lipkin, 1971). After the period of maturation, the cells leave the proliferating pool and enter the compartment of the nondividing cells, keeping the previously incorporated radioactive nuclear labeling. Thus, these cells are labeled though they are no longer able to synthesize DNA (e.g., orthochromatic erythroblasts, metamyelocytes, granulocytes). From the increase in the labeling indices conclusions regarding the production rate of the cell populations can be drawn (Fliedner et al., 1959, 1961). After another period of time, which is also de-

3

pendent on the turnover of the cells, the labeling indices decrease in intensity. Now the injected radioactivity has been diluted by cell division and finally is no longer demonstrable. Besides, primarily nonlabeled cells enter and pass the cell cycle to a growing extent.

In these long-term experiments the reincorporation of labeled DNA compounds from nuclei of decayed cells is of special importance. Such a reutilization has been described in a number of physiological and pathological processes (Bryant, 1963; Gerecke and Gross, 1976; Helpap et al., 1971; Helpap and Cremer, 1972). Reutilization phenomena are to be expected in hematopoietic organs (bone marrow, spleen) and have been proved. The probability of such a process is already suggested by the observation of ^{3}H-TdR-labeled expelled erythroblast nuclei taken up by macrophages (Cottier et al., 1963). According to studies of Heininger et al. (1971) and Feinendegen et al. (1966), around 50%–60% of the ^{3}H-TdR incorporated by "blasts" originates from the nuclei of decayed cells. The rate of reutilization is around 35% in the bone marrow of rats and around 38% in the spleen of mice (Feinendegen et al., 1973). An increased rate of reutilization in granulocytopoiesis can be expected only 72 h after injection of the isotope (Gerecke and Gross, 1976).

In autoradiographic investigations the extent of the "background" should be known. Background means the amount of silver grains not produced by the injected tritium, but by other causes such as cosmic radiation, mechanical or electrostatical influences (Baserga and Malamud, 1969). Therefore corrections in the determination of the labeling index and the density of the silver grains may be necessary to allow the comparison of these parameters among different investigators. However, the overall configuration of the curves remains unchanged (Fliedner et al., 1961).

2.1 Semithin-Section Autoradiography

Usually histologic sections of paraffin-embedded material, 3–5μm in thickness, are used for autoradiography. Generally, stripping films yield quite good results. But the light-optical resolution in paraffin-section autoradiographs is not high enough to recognize smaller cell structures and to identify different hematopoietic cell elements. Autoradiographs of blood, bone marrow, and spleen smears allow a definite differentiation of the blood cells and their precursors; but they do not provide any information about the topographic origin of the detached cells from various anatomic structures. Besides, in smears a certain selection of the cells may occur, since less adherent cell populations come off more easily than more fixed ones.

For the above-mentioned reasons the semithin-section autoradiography (Blümcke and Backmann, 1966; Yamashita and Helpap, 1974a) is a very suitable method for the quantitative autoradiography of hematopoietic organs. The topography of the cell systems remains intact and cell differentiation is possible. Because of their high light-optical resolution, semithin-section autoradiographs have a fixed position between light- and electronoptical autoradiography (Amlacher, 1974). Compared to the smear technique, there is the disadvantage of a lower number of cells that can be evaluated. Therefore, several semithin sections are required from different organ sites.

Preparing semithin-section autoradiographs, the tissue is fixed in suitable solutions, embedded in methylacrylate, Araldite, or Epon, and cut with the ultramicrotome. The sections with a thickness of 0.8–1.0 μm remain in the embedding material where they keep their even surface. They are coated with a photographic emulsion. In staining with di- or polychromatic stains the following points have to be considered:

1. The absorption of the dye by the photographic layer
2. The dislocation or destruction of the photographic layer during the process of staining
3. The selection of suitable stains; this is limited according to the embedding medium used

The di- or polychromatic staining methods described by North (1971) and by Yamashita and Helpap (1974a) — which were applied in the present study — have proved useful for the semithin-section autoradiography of lymphatic and hematopoietic tissue.

3 Material and Methods

3.1 Development of Spleen and Body Weight

The following investigations were carried out on Wistar rats (strain Hannover) of both sexes with known dates of birth. The animals were kept under standard, non-germ-free conditions in the laboratory of the Institute of Pathology the University of Bonn and fed on Altromin and water ad libitum.

Spleen and body weight of 131 rats of the following age groups were recorded (the number of animals appearing within the brackets): 1 h (10), 18 h (13), 3 d (11), 5 d (9), 10 d (8), 14 d (10), 20 d (11), 30 d (22), 45 d (21), 60 d (16).

3.2 Autoradiographic Investigations on the Red Pulp of Rats of Different Age Groups

One hour prior to death 38 of the above-listed animals were administered an intraperitoneal injection of 3 μCi ^{3}H-thymidine per gm body weight (NET -o27X; New England Nuclear Corporation, Boston, Mass. USA). The specific activity was 20 Ci/mMol. The distribution of the animals on the various time intervals was the following: 1 h: five animals; 18 h: 3; 3 d: 2; 5 d: 3; 10 d: 2; 14 d: 4; 20 d: 5; 30 d: 6; 45 d: 4; 60 d: 4.

The animals were killed by decapitation: from the 5th day of life onward they received a slight ether anesthesia.

Evaluation of the Histologic and Autoradiographic Preparations. The various hematopoietic and stromal cells in the red pulp were classified according to nuclear size, nuclear shape, nuclear structure (distribution of chromatin, nucleoli), and amount and staining of cytoplasm (Weicker, 1954a; Lennert, 1952; Yamashita and Helpap, 1974b; Bessis, 1973; 1977).

The cells in sections differ from those in smears in the following ways (Lennert, 1952; Leibetseder, 1948):

1. Due to fixation and embedding the cells in sections are smaller by a third than those in dry smears.
2. In the section the cells exhibit transparent vesicular nuclei with some nuclear clumps and fine chromatin filaments. In the smear, however, the nuclei appear much denser and darker.

In the semithin sections an analysis of the various hematopoietic cells and the cells of the supporting tissue was carried out. The distribution in percentage and the cell number per unit area (UA = 0.01 mm^2) were computed. For this purpose an ocular with a counting eyepiece graticule providing squares of 100 μm length at a magnification of 1000 was used (Calvo et al.,

5

1975). As in quantitative investigations on the bone marrow, 2500–3000 cells/animal were analyzed (Fritsch and Queißer, 1978).

In the autoradiographs the percentage of the red pulp cells which were radioactively labeled was determined. The ^{3}H-TdR indices of all proliferating hematopoietic cells were calculated separately for different precursors. Fifty to 100 cells of morphologically clearly defined populations per animal, and in older age groups (45 and 60 days) around 50, were counted. In addition the mitotic indices of the overall erythropoiesis, granulocytopoiesis, and basophilic blasts were determined. For each group the ratio of the labeling to the mitotic index was plotted against the age, as outlined in the diagrams.

Finally, the labeling pattern of the stromal cells was investigated. For each experiment the mean value and corresponding standard error (SE) were entered into the diagram.

3.3 Cell Kinetic Studies on the Red Pulp of Newborn Rats

This study was done on 33 Wistar rats of different breeds. Sexes were not determined. One hour after birth each animal received an intraperitoneal injection of 3 μCi ^{3}H-TdR/g body weight. The rats were killed after various time intervals: 1 h (5), 6 h (2), 12 h (3), 18 h (2), 21 h (2), 44 h (3), 3 d (1), 5 d (1), 7 d (3), 10 d (3), 12 d (3), 14 d (3), and 28 d (2).

Evaluation of the Autoradiographs. The following parameters were determined:
1. The percentage of the radioactively labeled interphase nuclei of the red pulp cells
2. The labeling indices of large and small basophilic blasts
3. The percentage of radioactively labeled erythropoietic cell elements, the labeling indices of basophilic, polychromatic and orthochromatic erythroblasts and the number of radioactively labeled expelled free and phagocytized erythroblast nuclei
4. The percentage of radioactively labeled cells of the proliferating granulopoiesis (promyelocytes and myelocytes) and the labeling indices of metamyelocytes and granulocytes
5. The labeling index of lymphocytes in the red pulp
6. The labeling indices of megakaryocytes, phagocytizing reticulum cells (macrophages), non-phagocytizing reticulum cells, and endothelial cells

At least 80–100 cells were counted in one cell group.

3.4 Autoradiographic and Histologic Technique

The spleen of the killed animal was removed and cut into halves longitudinally. One half was fixed in a 6% solution of formalin with 0.5% of trichloroacetic acid and 0.1 mg/ml inactive thymidine (4 °C, 24 h). The fixed tissue was washed in running tap water for 24 h. After the passage through increasing concentrations of ethanol it was embedded in paraffin. Then histologic sections of 3 μm thickness were prepared and placed on glass slides coated for better adherence with a gelatin film. After deparaffinization in xylene and the passage through decreasing concentrations of ethanol, the sections were placed into distilled water. The preparations were then coated with stripping AR 10 films (Kodak) under redlight in the dark room (Kodak Wratten filter 1). The autoradiographs were stained with hemalaun through the photographic emulsion. On other sections the naphthyl-ASD chloroacetic esterase reaction was done before film coating (Leder, 1967).

The paraffin sections meant for ordinary histologic examination were stained with hematoxylin-eosin and alcian blue, after Giemsa and Gordon-Sweet. The PAS reaction was also performed.

From the second part of the spleen 1 mm slices (8–10) were fixed in 2% OsO$_4$ and phosphate-buffered 2.5% glutaraldehyde. The specimens were dehydrated in increasing concentrations of alcohol and embedded in Epon (Luft, 1961).

Preparation and Staining of the Semithin-Section Autoradiographs. From the tissue pieces embedded in Epon 0.8–1.0 μm semithin sections were cut with the LKB III ultramicrotome. By means of small water droplets they were placed on clean glass slides wetted beforehand with a solution of gelatin and chrom alum. Then the slides were stretched on a hot plate at a temperature of 70 °C.

The coating with a liquid photographic emulsion (Kodak NTB 2) according to the dipping technique was done under red light in the darkroom.

First the photographic emulsion was diluted 1:1 with distilled water and liquified in the water bath at 40 °C. The slides were dipped once into this diluted emulsion and the backside was immediately wiped with filter paper. After drying they were stored, together with $CaCl_2$ as a hygroscopic agent, in light-tight boxes in the darkroom at 4 °C.

The exposure time was 21 days. The photographic development of paraffin-section and semithin section autoradiographs was done in the conventional manner (Schultze, 1968).

The staining of the film-coated and noncoated semithin sections was performed according to the method of Yamashita and Helpap (1974a). Some of the sections were stained only with methylene blue (Richardson et al., 1960).

In control sections the "background fog" was constantly around 2–3 silver grains, thus nuclei with 4 or more silver grains were regarded as labeled.

4 Cytology and Definition of the Rep Pulp Cells

Basophilic Blasts

Two types can be distinguished: large basophilic blasts (more than 10 μm in diameter) and small basophilic blasts (less than 10 μm in diameter). Both have pale, large, roundish to oval, relatively plump nuclei. The nucleoli stain violet; they are located partly in the center, partly at the nuclear membrane. The cytoplasm is clearly recognizable and strongly basophilic; sometimes it contains small vacuoles (Figs. 1, 2).

Erythroblasts

Three types can be distinguished in the semithin section (five in the smear). The macroblasts were enclosed within the group of basophilic erythroblasts and the proerythroblasts could not be differentiated.

Basophilic Erythroblasts (Fig. 3a, b). The size of these round cells is 7–9 μm. They have a dark nucleus with more or less regularly distributed dense chromatin clots. Sometimes small nucleoli are present. The nucleus is surrounded by a narrow basophilic cytoplasmic rim.

Polychromatic Erythroblasts (Fig. 3c, d). These cells with a size of approximately 5–7 μm stand in between the basophilic and orthochromatic erythroblasts. The dark nucleus exhibits a chessboard-like structure of the chromatin, due to dense and regularly distributed chromatin clots. There is more cytoplasm; because of a more intensive hemoglobin production it is of a grey-violet to reddish-pink color.

Orthochromatic Erythroblasts (Fig. 3e). In the semithin sections the size of these cells is around 4–5 μm. The nucleus is small and the ratio nucleus/cytoplasm even more in favor of the cytoplasm. Because of the condensed chromatin structure the nucleus looks dark, sometimes still suggesting a chessboard pattern; nucleoli cannot be recognized. Often the nucleus is localized eccentrically in the clearly acidophilic dark-red stained cytoplasm.

Expelled Erythroblast Nuclei (Fig. 3e). These small dark pyknotic nuclei were expelled from the erythroblasts at the end of their maturation. Partly they lie freely in the

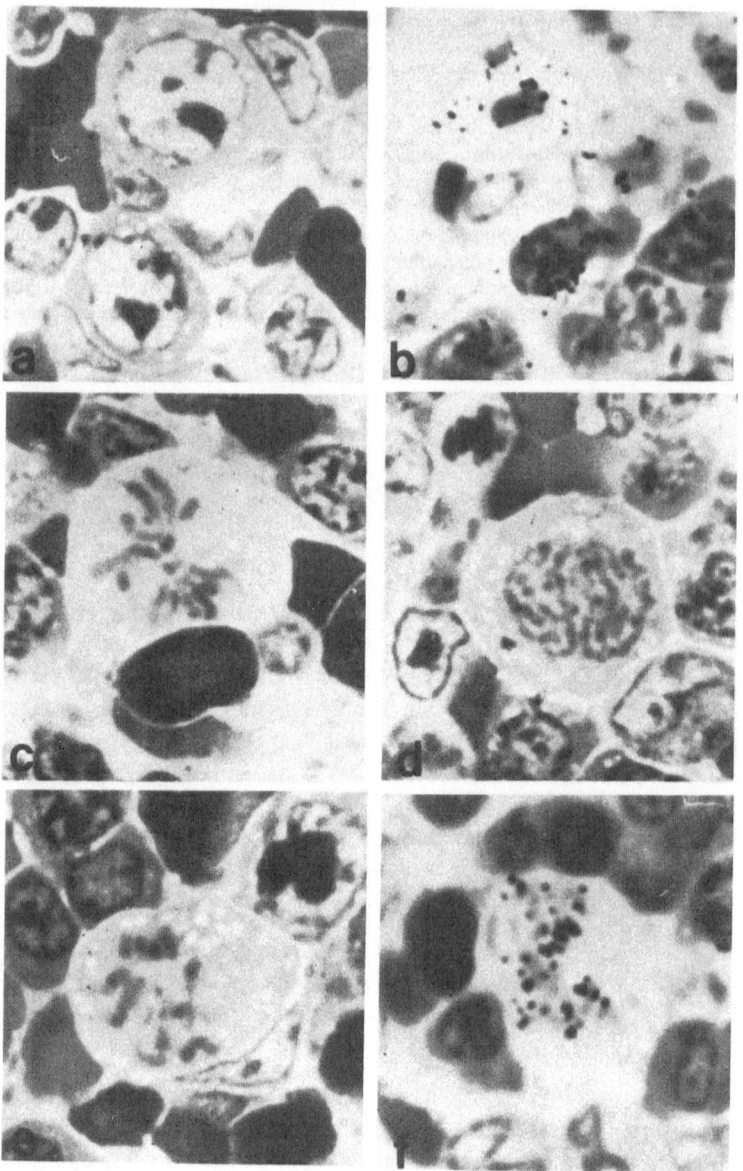

Fig. 1a–f. Nonlabeled (*a*) and labeled (*b*) large basophilic blasts and mitoses (*c–f*). × 1250

sinuses, partly they have already been phagocytized by reticulum cells. The number of these naked nuclei can be taken as a relative measure of the reticulocyte production.

Myelopoietic Cells

Immature myelopoietic cells comprise promyelocytes and myelocytes, which are still capable of dividing, and metamyelocytes. Myeloblasts could not be identified in the

8

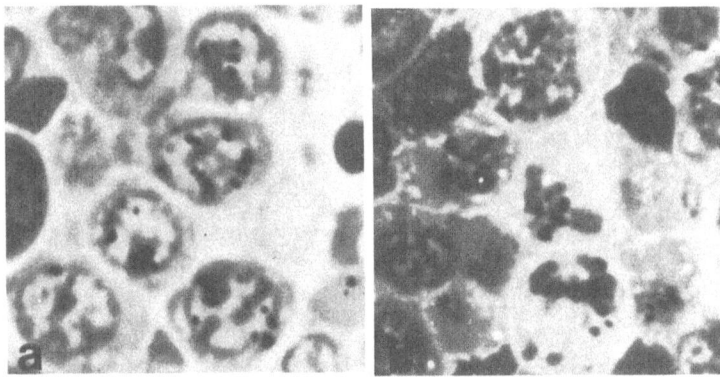

Fig. 2a, b. Small basophilic blasts (*a*) and mitoses (*b*). × 1250

semithin sections. Eosinophilic granulocytes and their precursor cells were not differentiated.

Promyelocytes. This is the largest and most immature myelopoietic cell which is clearly recognizable in the semithin sections (Fig. 4a–c). It is about 10–12 μm large and has an oval, slightly concave, eccentrically located nucleus. This is rather pale; at the nuclear membrane small chromatin condensations are apparent. Nucleoli, 2–4 μm in size, can be found quite regularly. Only a few cells show a ring-shaped nucleus. The cytoplasm is sharply defined and mainly basophilic; however, small perinuclear acidophilic areas are already visible.

Myelocytes (Fig. 4d, e). These are cells measuring 6–9 μm, with an ovaloid, reniform dark nucleus; nucleoli are not visible. Many of these cells show a ring-shaped nucleus, typical of the granulopoiesis in rodents. The cytoplasm has an acidophilic red color. Granules are not or only faintly visible against the background of the dark cytoplasm.

Metamyelocytes (Fig. 4f, g, h). They measure about 6 μm and have a deeply indented reniform nucleus. The nuclear chromatin is condensed at the margins forming a distinctly outlined nuclear membrane. Occasionally, small chromatin clots can be observed at both nuclear poles. Again, typical ring-shaped nuclei are present. The cytoplasm shows an intensely red color.

Polymorphonuclear Granulocytes (Fig. 4i). They are 4–5 μm in size and exhibit the well-known segmented nuclear configuration. Stab cells were not considered separately. The cytoplasm stains acidophilic red; granules are not visible.

Lymphocytes

Lymphocytes with a diameter of 3.5–7.0 μm appear as characteristic cells in the semithin section. The narrow cytoplasmic rim is generally roundish to oval shaped; the nucleus occupies nearly all of the cell, having a spherical configuration, with the chromatin near the nuclear membrane being irregularly condensed. In most cases a centrally

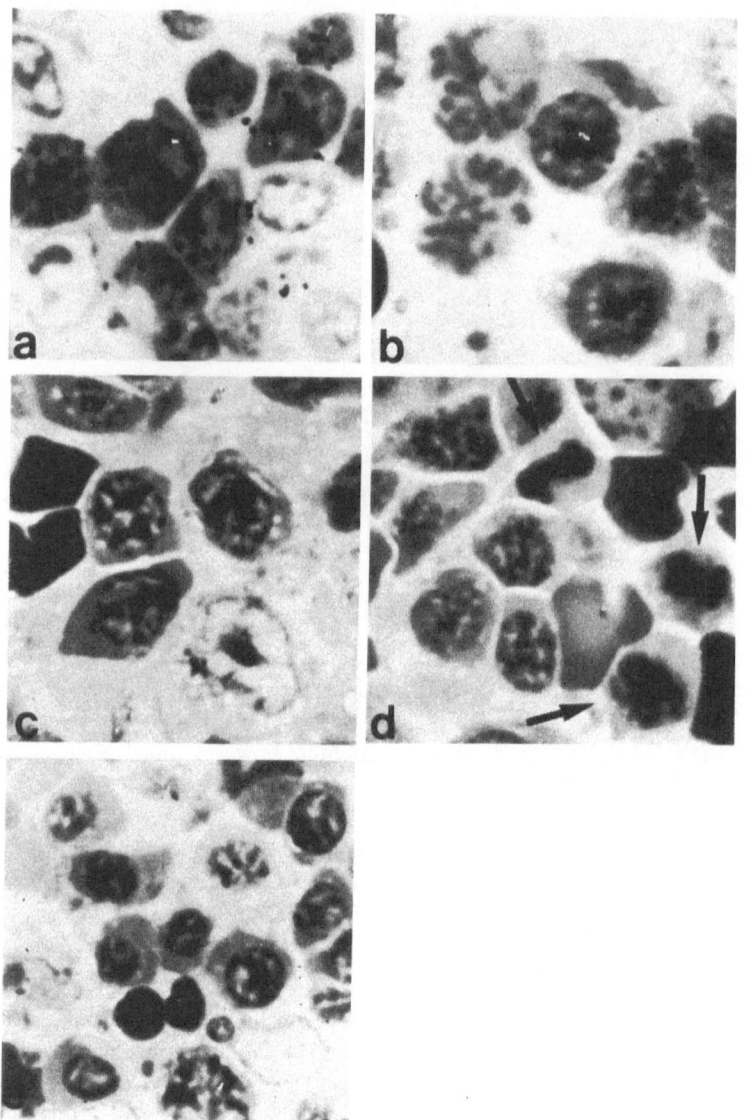

Fig. 3a–e. Erythropoiesis in the rat spleen: (*a, b*) basophilic erythroblasts and mitoses; (*c, d*) polychromatic erythroblasts and mitoses; (*e*) orthochromatic erythroblasts, expelled erythroblast nuclei. × 1250

located nucleolus can be observed (Fig. 5). The lymphocytes were not differentiated into small, medium, and large ones.

Reticulum Cells

According to their functional state they were distributed into phagocytizing (macrophages) and nonphagocytizing ones.

10

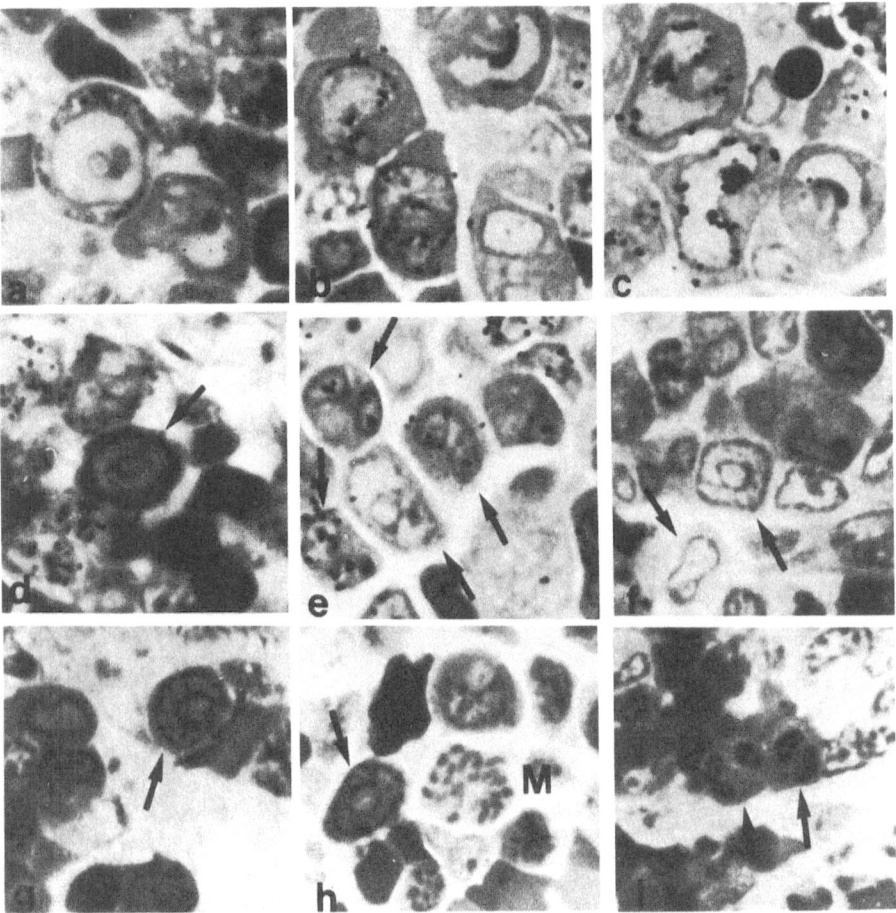

Fig. 4a–i. Granulocytopoiesis in the rat spleen: (*a, b, c*) radioactively labeled and nonlabeled promyelocytes; (*d, e*) labeled and nonlabeled myelocytes; (*f, g, h*) metamyelocytes and myeloic mitosis (*M*); (*i*) polymorphonuclear granulocytes. × 1075

Nonphagocytizing Reticulum Cells. They are of different sizes; the cytoplasmic outlines cannot be determined in the semithin sections. The nuclei are rather pale and poor in chromatin. They have a longish to stellate, oval or fusiform shape and an irregularly outlined nuclear membrane. The nuclei are of various sizes. Generally one or two large nucleoli are visible (Fig. 6).

Phagocytizing Reticulum Cells (Macrophages). Besides the above-mentioned criteria, these cells exhibit phagocytized material (nuclear and cellular debris; Fig. 7). Often they have ingested expelled erythroblast nuclei. Hence the reticulum cells in the center of erythroblastic islands ("nurse cells") belong to this cell group.

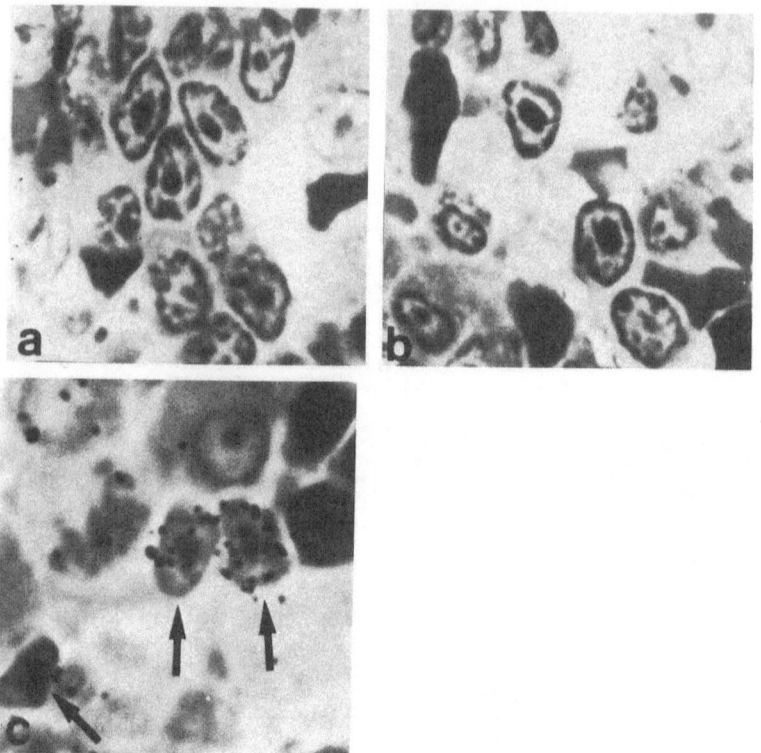

Fig. 5. Nonlabeled and labeled (44 h after pulse-labeling with ^{3}H-TdR) lymphocytes in the red pulp. × 1250

Monocytoid Cells

This term applies to cells with a size of 6–10 μm in the semithin section. The nuclear membrane shows characteristic bizarre indentations. Occasionally a small, inconspicuous nucleolus is found. The cytoplasm is fairly well recognizable and stains bluish-gray (Fig. 8).

Endothelial Cells

Endothelial cells (Fig. 9) are the lining cells of the sinusoids, capillaries, and venules. They have an oval to longish shape, a spindle-shaped nucleus and large nucleoli. The nuclear chromatin is loosely arranged; the cellular membrane sharply defined. These sinusoidal endothelial cells could not always be distinguished from fibroblastic cell elements with certainty. Therefore, a chain-like arrangement was regarded as an indication of their endothelial nature.

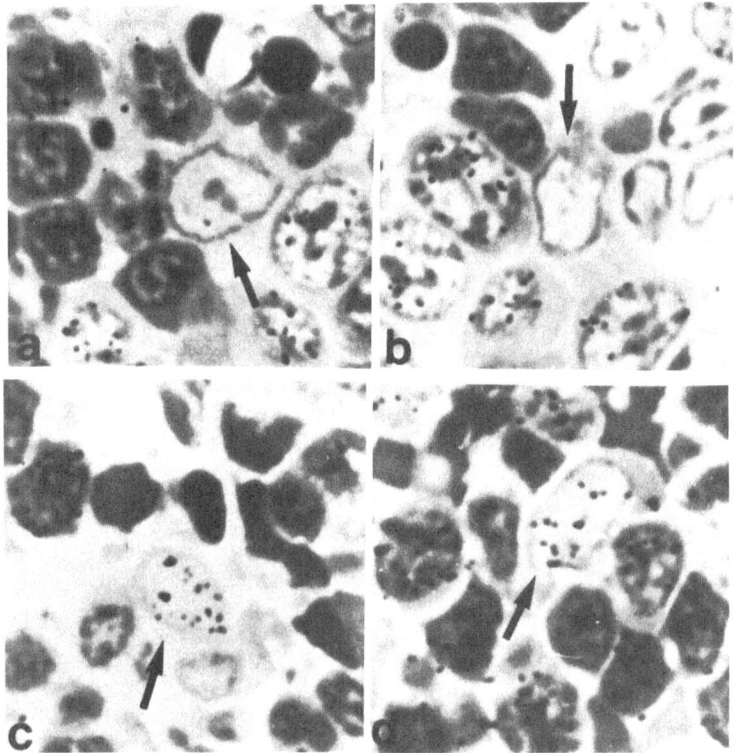

Fig. 6. Nonlabeled and labeled reticulum cells without phagocytosis. × 1250

Megakaryocytes

The size of these multinuclear giant cells varies in the semithin section between 10–35 μm depending on nuclear ploidy. Differences in nuclear and cytoplasmic structures allow distinction to be made between the following maturation stages (Fig. 10):

Megakaryocyte Type I (Megakaryoblast). It contains one or more trapezoid to longish-oval nuclei with two or three nucleoli. The cytoplasm is free of granules and is homogeneous basophilic.

Megakaryocyte Type IIa (Basophilic Megakaryocyte). This cell, too, possesses several nuclei with distinct nucleoli. The central parts of the cytoplasm contain acidophilic granules, whereas the peripheral parts appear homogeneous basophilic.

Megakaryocyte Type IIb (Granulated, Thrombocyte-producing Megakaryocyte). The nuclear chromatin has a fairly dense structure, the nucleoli are smaller. The acidophilic cytoplasm is rich in granular structures separated by so-called demarcation lines.

Megakaryocyte Type III. There are only small irregular cytoplasmic rims visible. The nuclei have undergone pyknosis and appear dark and dense.

The results regarding cell distribution and proliferation refer to the total of all megakaryocytes.

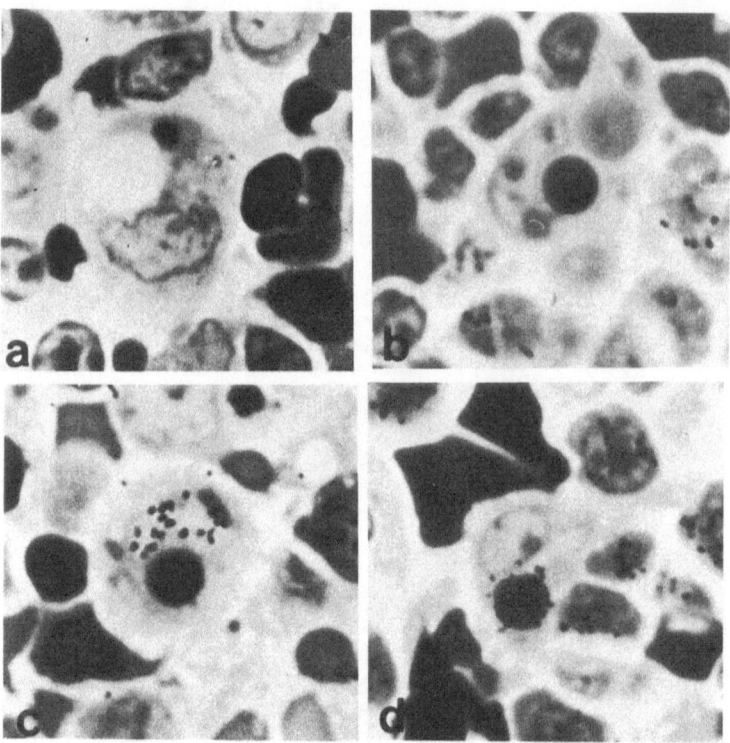

Fig. 7. Labeled and nonlabeled macrophages; partly after incorporation of ^{3}H-TdR-labeled erythroblast nuclei. × 1250

Plasma Cells

Plasma cells (Fig. 11) exhibit an eccentrically located large nucleus with the well-known cartwheel structure. The abundant and clearly recognizable cytoplasm has a bluish-gray color.

5 Results

5.1 The Anatomic Development of the Postnatal Spleen

5.1.1 The Development of Spleen and Body Weight

The spleen weight of newborn Wistar rats fluctuates between 7 and 8 mg, and the body weight is around 6 g. During the following days and weeks a continuous acceleration in the weight gain of both body and spleen weight sets in, making the curves run parallel. The steepest rise occurs between the 20th and 30th day of life. However, the relative spleen weight, i.e., the percentage of the body weight, increases much

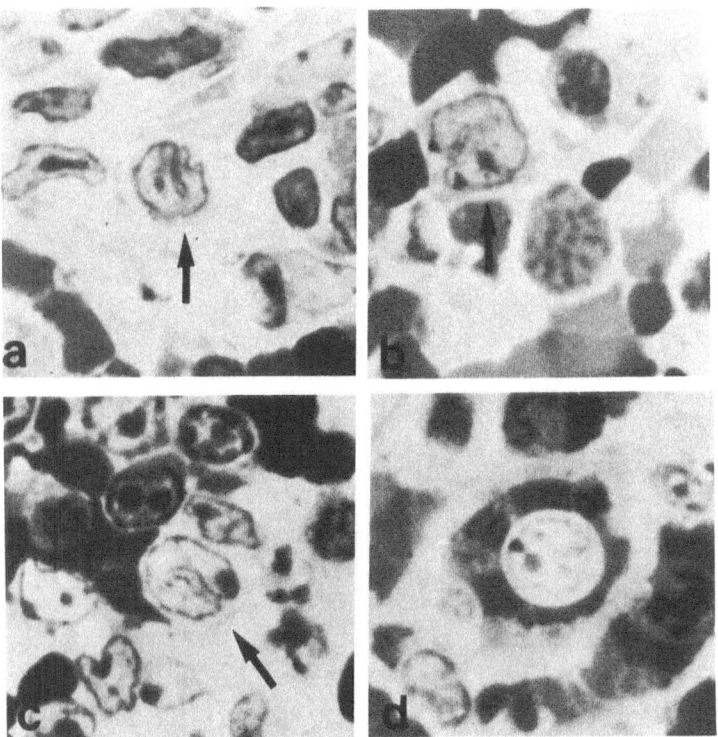

Fig. 8. Different types of "monocytoid" cells. × 1250

more steeply between 1 h and 5 days after birth. Between the 10th and 30th day the weight development is equal. Thereafter the body weight increases comparatively more, lowering the ratio of spleen to body weight again (Fig. 12).

5.1.2 The Histologic Development of the Red Pulp

The spleen of newborn rats is rich in immature hematopoietic, predominantly erythropoietic cells. There are many mitoses. Similar to the erythroblast aggregations in the bone marrow, the immature erythropoietic cell elements are often located around a central reticulum cell. Numerous pyknotic, expelled free or already phagocytized erythroblast nuclei are visible. Occasionally, more mature erythroblasts can be seen in the lumina of small vessels. In between, single, mainly immature, granulopoietic cells are found, mostly situated by themselves or in pairs. They can be clearly distinguished by their intensely red cytoplasm in the naphthyl-ASD-chloroacetic esterase reaction. The immature reticular framework and the immature sinusoidal system are less prominent. Some arterioles can be observed surrounded by a few monocytoid and lymphoid cells.

From the 3rd day of life onward, focal myelopoietic cell aggregations in the otherwise unchanged red pulp turn up, consisting partly of more mature precursor cells.

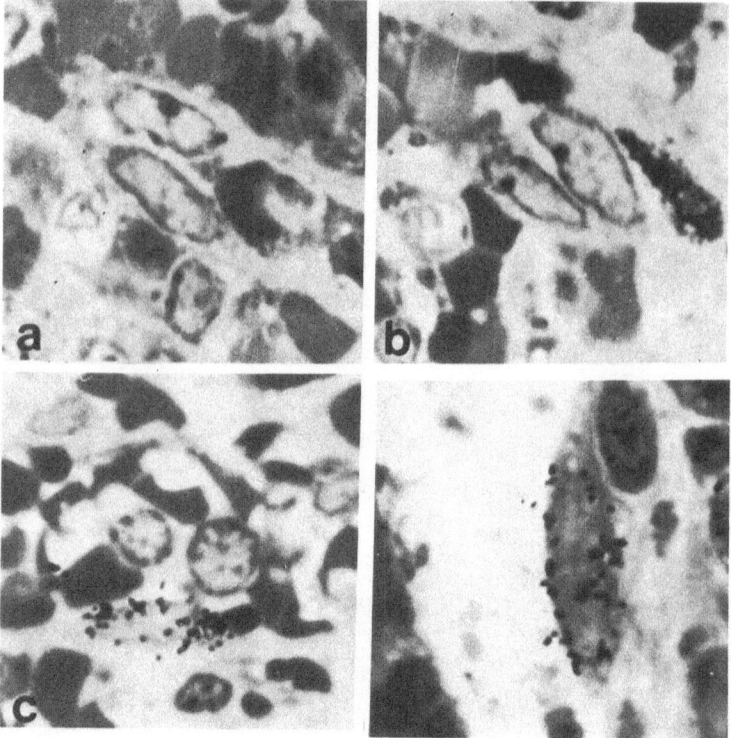

Fig. 9. Labeled and nonlabeled endothelial cells. × 1250

This hematopoietic activity is also obvious on the 5th and 10th day. After the 5th day, however, narrow trabecular-like fibrous bands are found. Now the granulo-cytopoiesis is obviously located preferably in the immediate surroundings of these areas. This becomes more evident between the 10th and 14th day, when the sinus-oidal system and the trabecular structure stand out more markedly. At the same time there more white pulp areas are visible, and therefore the red pulp can be sub-divided into a subcapsular and an interfollicular region. After 14 days the red pulp exhibits the typical mature structure of a cavernous system, connecting arterial and venous vessels and consisting of vessel-like sinuses and surrounding cavernous pulp.

Until the 45th day erythropoietic "basophilic islands" are still clearly visible, however, to a decreasing extent. There are only a few groups of loosely arranged myelopoietic cells, sometimes in a subcapsular or peritrabecular localization. By the 60th day the hematopoietic activity has further diminished but has not yet comple-tely ceased.

5.2 Analysis of Hematopoietic and Stromal Cells in the Red Pulp of Differently Aged Rats

The following curves illustrate the cell number/0.01 mm² spleen area; the corres-ponding percental distributions are listed in tables. The latter will serve for compari-

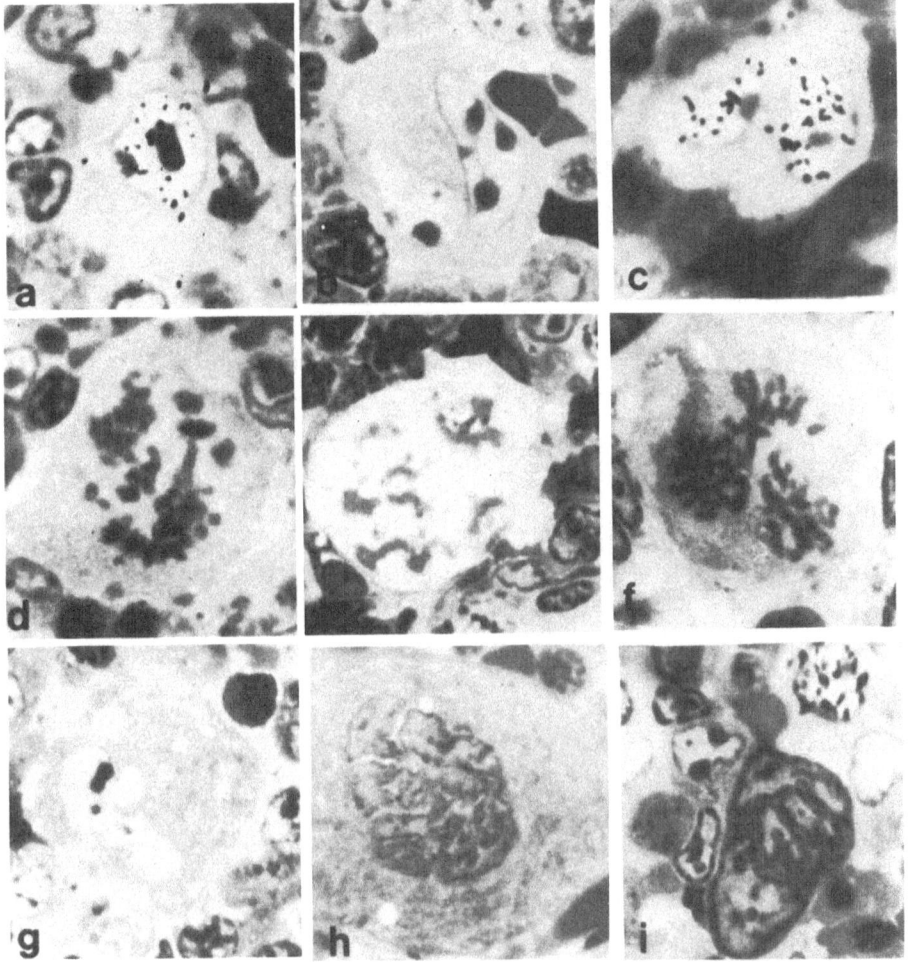

Fig. 10a–i. Megakaryocytopoiesis in the rat spleen: (*a*) labeled large basophilic blast; (*b*, *c*) nonlabeled and labeled megakaryocyte type I; (*d*, *e*, *f*) different mitotic figures of type I megakaryocytes; (*g*) megakaryocyte type IIa; (*h*) more mature megakaryocyte type IIb; (*i*) megakaryocyte type III. × 1000

son with splenograms by other investigators. On the whole, values in the tables and curves agree. However, values in the tables are subjected more to changes in the frequency distributions of single cell populations, thus not giving the true pattern of distribution. Such eventual alterations do not affect the number of cells expressed in cell number/unit area, since the unit area (UA = 0.01 mm^2) is a fixed parameter (Calvo et al., 1975; Ruhrmann, 1966).

In Fig. 13 the cell concentration of the total red pulp is plotted against the age. The cell number increases from approximately 39 cells/UA to a maximum of 63/UA on the 10th day. Thereafter a continuous diminution occurs until the 60th day, when the cell number is the same as at birth, i.e., 39 cells/UA.

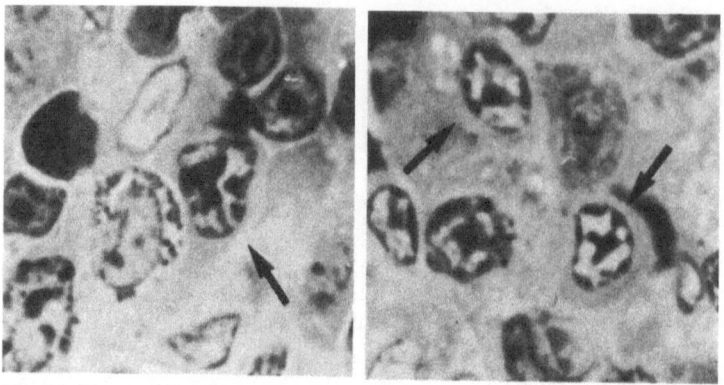

Fig. 11. Plasma cells in the red pulp. × 1250

Small Basophilic Blasts

As shown in Fig. 14, 1 h after birth four small basophilic blasts/UA are found in the red pulp. Their number doubles by the 5th day (8.6/UA). Thereafter the curve declines, with a steeper fall at the beginning, until the 60th day when only 0.9 cells/UA are present. Table 1 shows the distributions of small and large basophilic blasts in terms of percentage.

Large Basophilic Blasts

The number of large basophilic blasts reaches its maximum between 18 and 72 h post-natally. By then the cell concentration has increased to 5.3 (from 2.2 at birth). Then

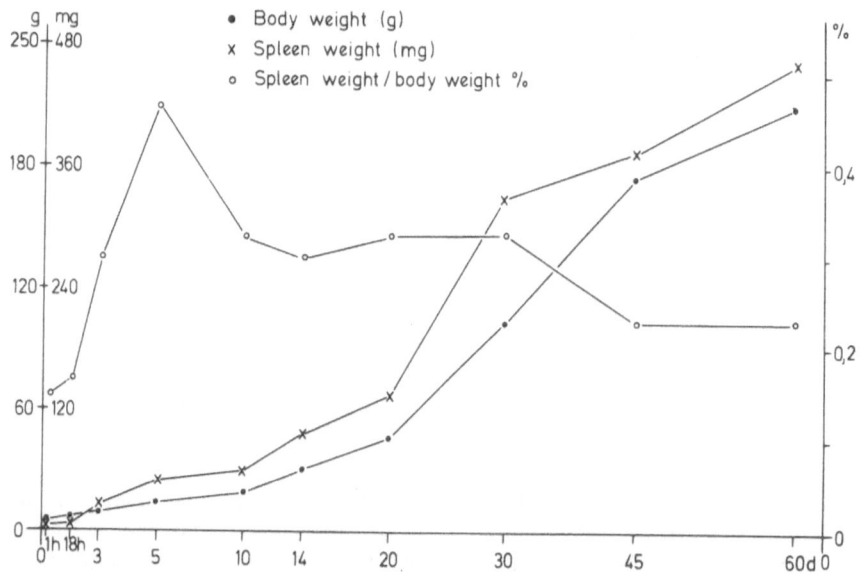

Fig. 12. The development of spleen and body weight in growing rats

18

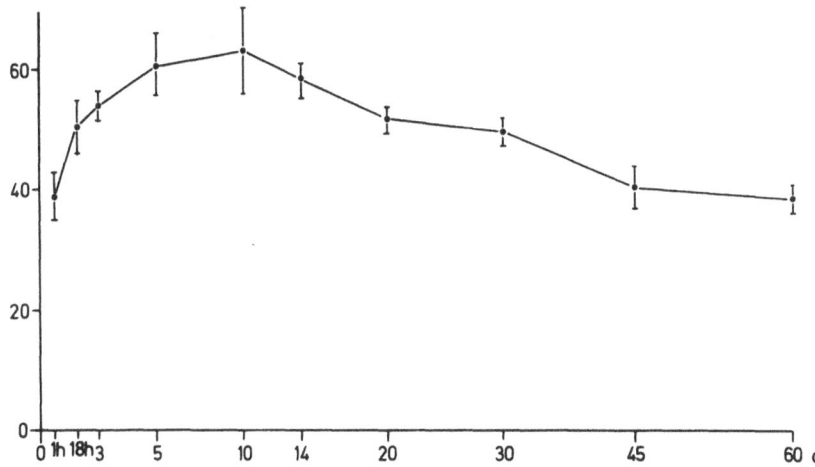

Fig. 13. Total cell number/unit area in the red pulp of growing rats

Table 1. Percentage of small and large basophilic blasts in the red pulp in correlation with age

Age	Small basophilic blasts[a]	Large basophilic blasts[a]
1 h	10.7 ± 2.3	5.5 ± 1.9
18 h	11.6 ± 1.2	10.3 ± 1.2
3 d	15.8 ± 4.9	9.2 ± 3.5
5 d	13.9 ± 0.8	5.2 ± 0.2
10 d	5.8 ± 1.1	3.0 ± 0.3
14 d	5.5 ± 0.9	1.7 ± 0.4
20 d	4.1 ± 0.6	1.6 ± 0.4
30 d	6.9 ± 1.7	2.3 ± 0.7
45 d	3.8 ± 1.0	1.4 ± 0.5
60 d	2.2 ± 0.6	0.7 ± 0.6

[a] Mean ± standard error

the curve falls gradually; between the 14th and 45th day of life the concentration of the large basophilic blasts remains at about one cell/UA (Fig. 14; Table 1).

Total Erythropoiesis

Erythroblasts

The concentration of erythroblasts in the red pulp increases steadily (by the factor 3) until a maximum of 35 erythroblasts/UA is reached on the 10th day. Their number drops markedly between the 14th and 30th day of life; after 8–9 weeks only a few are left (1.7/UA; Fig. 15; Table 2).

Figure 16 illustrates the quantitative relationship between the different stages of erythropoiesis at various time intervals after birth. (The mean value of animals examin-

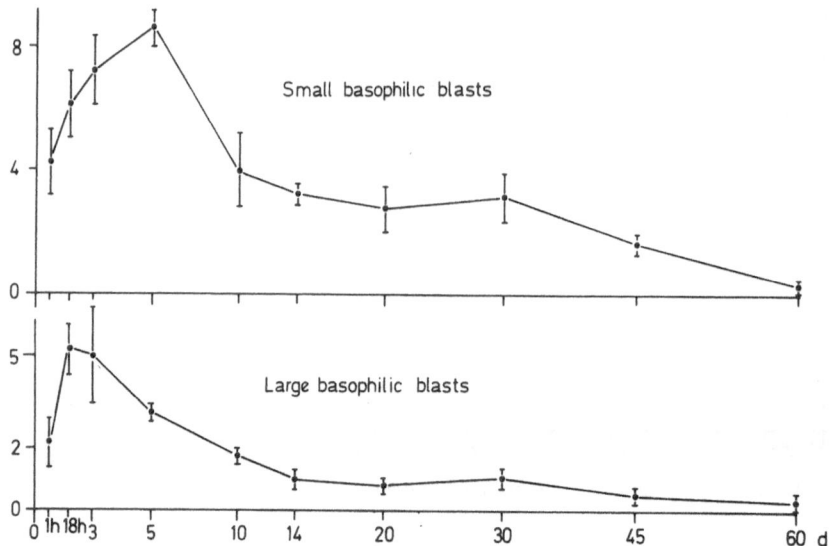

Fig. 14. Number of small and large basophilic blasts/unit area in the red pulp of growing rats

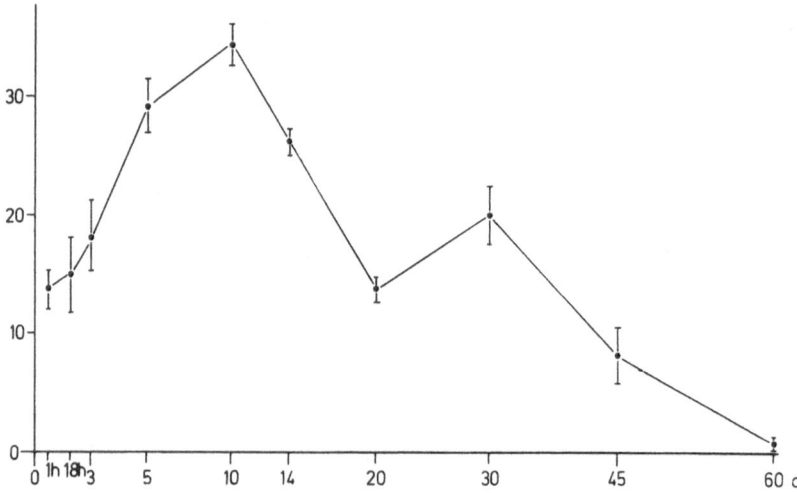

Fig. 15. Number of erythroblasts/unit area in the red pulp of growing rats

ed at the same age was entered into the diagram.) The percentage of basophilic erythroblasts increases slightly until the 18th h after birth; it then decreases to a minimum on the 14th day. Thereafter the amount remains around 16%. At the beginning the percentage of the polychromatic erythroblasts is twice, later on even three times as high as that of the basophilic erythroblasts. The percentage of orthochromatic erythroblasts comes close to that of polychromatic erythroblasts, while later it becomes markedly less (20, 30, 60 d).

20

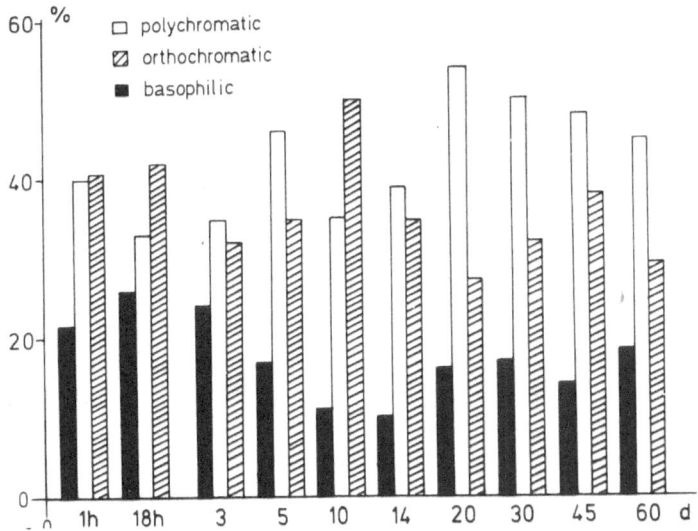

Fig. 16. Relationship between different erythroblasts (in percentage) in the red pulp of growing rats

Table 2. Percentage of erythropoietic cells in the red pulp in correlation with age

Age	Erythropoiesis[a]
1 h	32.5 ± 4.0
18 h	27.2 ± 4.6
3 d	36.3 ± 5.3
5 d	46.2 ± 3.5
10 d	48.0 ± 2.4
14 d	44.3 ± 2.7
20 d	26.7 ± 4.6
30 d	39.1 ± 2.5
45 d	18.6 ± 5.1
60 d	4.4 ± 1.4

[a] Mean ± standard error

These trends in the ratio among the immature cells of erythropoiesis become more obvious if the mean values of three longer time intervals are considered: birth–5 days, 10–20 days, 30–60 days (Table 3).

By the 5th day out of 100 immature erythropoietic cells roughly 20 are basophilic, 40 polychromatic, and 40 orthochromatic. Between the 10th and 20th day the ratio is 16:44:40, and between the 30th and 60th day it is 16:49:35.

These figures demonstrate that the percentage of the polychromatic erythroblasts is first twice, then three times as high as that of the basophilic erythroblasts. Such an increase cannot be observed in the ratio of the orthochromatic to the polychromatic erythroblasts.

Table 3. Quantitative relationship between the different erythroblasts from various age groups

Age	Basophilic erythroblasts	Polychromatic erythroblasts	Orthochromatic erythroblasts
0.– 5th d	19.7%	40.9%	39.4%
10.–20th d	16.3%	44.2%	39.5%
30.–60th d	16.3%	49.4%	34.3%

Erythroblastic Nuclei. Figure 17 shows the amount of free and phagocytized erythroblast nuclei in percentage relative to 100 erythroblasts. Immediately after birth a maximum of 12.4% is reached whereas after the 5th day only 5% are found. Until the 45th day the amount of erythroblast nuclei continues to decrease and fluctuates between 1.3% and 3%. On the 60th day expelled erythroblast nuclei in the red pulp have disappeared.

Myelopoiesis

The concentration of the granulocytes in the red pulp remains rather constant until the 5th postnatal day; by the 20th day it has increased by the factor 3.6. It then decreases to two cells/0.01 mm^2 and stays at this level until the 2nd month.

The maximum of the remaining granulopoietic cells (promyelocytes, myelocytes, metamyelocytes) can be observed after 18 h. It is followed by a decrease (factor 2) until the 5th day. A slight increase is visible on the 20th day. After 60 days immature granulopoietic cells have nearly disappeared (0.3 cells/UA; Fig. 18). Table 4 gives the percentage of mature and immature granulopoietic cell elements.

The ratio of immature proliferating myelopoietic cells (promyelocytes, myelocytes) to nondividing cells (metamyelocytes, granulocytes) was calculated (Fig. 19). Until the 3rd postnatal day cells of the immature granulopoiesis are obviously slightly dominant. From the 5th day onward the ratio is inversed. On the 60th day out of 100 granulopoietic cells in the spleen nine are immature and capable of DNA synthesis, while 91 are mature leukocytes circulating through the spleen or being retained in the spleen.

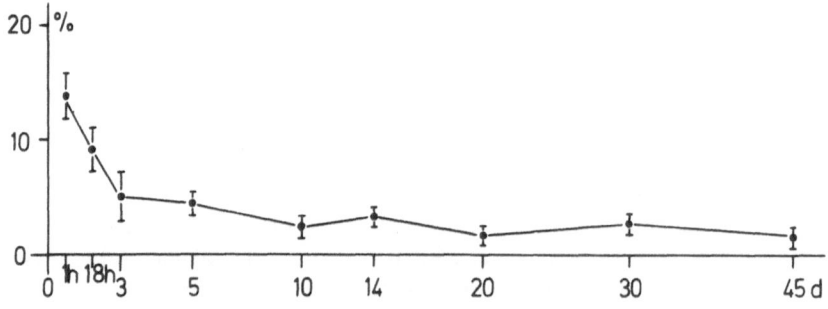

Fig. 17. Percentage of expelled erythroblast nuclei in the red pulp of growing rats

Granulocytes

Cell number/unit area

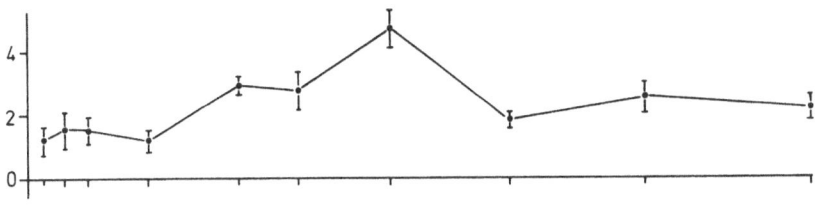

Immature granulocytopoiesis

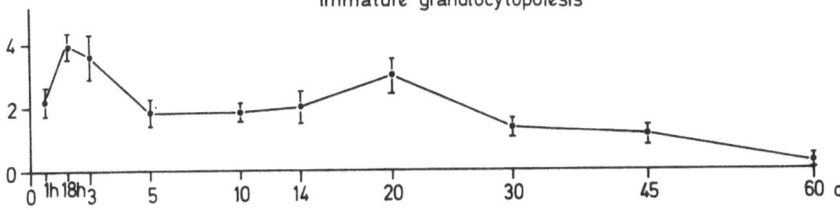

Fig. 18. Number of granulocytes and granulopoietic cells (promyelocytes, myelocytes, metamyelocytes) per unit area in the red pulp of growing rats

Lymphocytes

After an initial steep increase, the number of lymphocytes remains nearly constant until the 10th day of life (Fig. 20). Another steep rise then sets in which doubles the number of the cells again. Up until 60 days the curve remains at this level without remarkable changes. The percentile amount fluctuates around 6% until the 10th day and increases gradually from the 14th day onward (Table 5).

Plasma Cells

Plasma cells were observed only occasionally in the red pulp, from the 20th day onward.

Table 4. Percentage of granulopoietic cells (promyelocytes, myeolcytes, metamyelocytes) and granulocytes in the red pulp of growing rats

Age	Granulocytopoiesis (%)[a]	Granulocytes (%)[a]
1 h	6.3 ± 1.0	2.8 ± 0.6
18 h	8.1 ± 1.6	3.4 ± 1.2
3 d	6.4 ± 0.5	3.0 ± 0.6
5 d	3.2 ± 0.8	2.1 ± 0.3
10 d	2.5 ± 0.5	6.0 ± 1.8
14 d	4.7 ± 0.5	4.0 ± 0.7
20 d	5.4 ± 0.5	8.9 ± 1.1
30 d	3.2 ± 1.1	3.5 ± 0.3
45 d	2.8 ± 0.8	6.2 ± 0.9
60 d	0.8 ± 0.2	5.6 ± 1.1

[a] Mean ± standard error

23

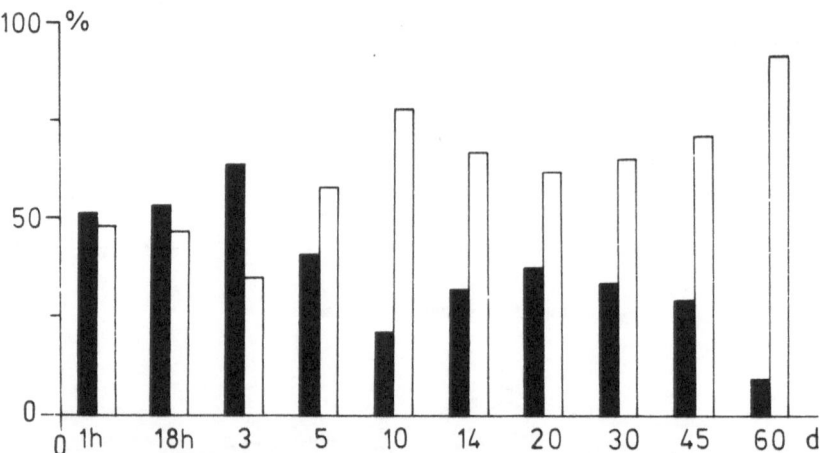

Fig. 19. Percentages of the dividing (■) granulopoiesis (promyelocytes, myelocytes) and non-dividing (□) granulopoiesis (metamyelocytes, granulocytes) in the red pulp of growing rats

Megakaryocytes

As outlined in Fig. 21, the concentration increases gradually by the factor 5 to 6 by the 14th day of life. Thereafter the number decreases continuously, until after 60 days the initial low value is reached again.

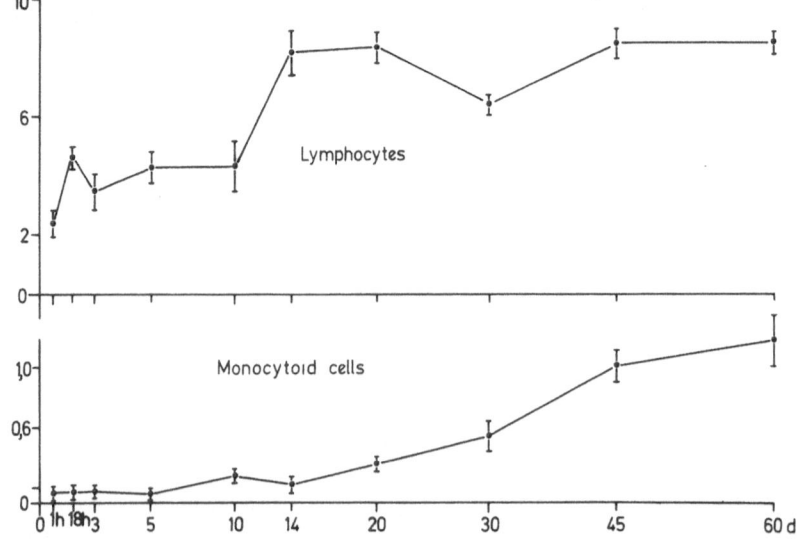

Fig. 20. Number of lymphocytes and monocytoid cells/unit area in the red pulp of growing rats

Table 5. Percentage of lymphocytes in the red pulp of growing rats

Age	Lymphocytes (%)[a]
1 h	5.5 ± 0.9
18 h	8.7 ± 0.6
3 d	6.9 ± 1.1
5 d	6.5 ± 1.2
10 d	5.8 ± 0.8
14 d	13.6 ± 1.0
20 d	14.7 ± 1.9
30 d	12.6 ± 1.0
45 d	16.9 ± 3.1
60 d	22.6 ± 0.6

[a] Mean ± standard error

Reticulum Cells

Non-phagocytizing Reticulum Cells. The concentration of this cell population remains remarkably constant throughout the experimental period. Disregarding smaller fluctuations, it is regularly around 8–11 cells/0.01 mm² spleen area (Fig. 22; percentages in Table 6).

Phagocytizing Reticulum Cells. Figure 22 also gives the number of macrophages per unit area. A definite increase appears only after the 45th day and on the 60th day the amount is twice as much as at birth.

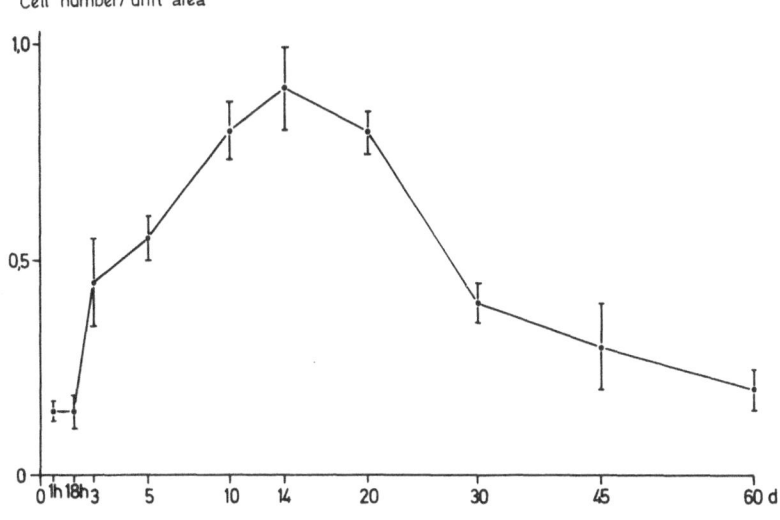

Fig. 21. Number of megakaryocytes/unit area in the red pulp of growing rats

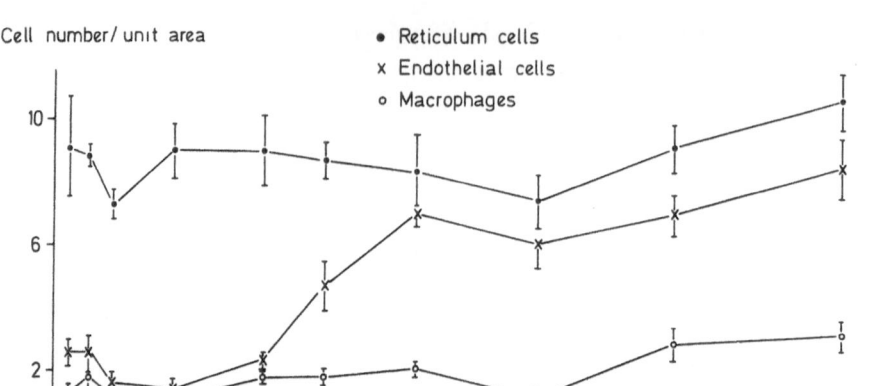

Fig. 22. Number of reticulum cells, endothelial cells and macrophages per unit area in the red pulp of growing rats

Endothelial Cells

After a steep increase between the 5th and 14th day the concentration of endothelial cells reaches its maximum on the 20th day (factor 3). Then the curve remains on this level; toward the end of the experimental period there is another slight ascent (Fig. 22, percentages in Table 6).

Monocytoid Cells

These are very infrequent cells in the perinatal rat spleen (Fig. 20, Table 6). After the 10th day there is an increase from 0.1 cell/UA to 1 cell/UA (60th day).

Table 6. Percentage of phagocytizing and nonphagocytizing reticulum cells, endothelial cells, and monocytoid cells in the red pulp

Age	Reticulum cells without phagocytosis[a]	Reticulum cells with phagocytosis[a]	Endothelial cells[a]	Monocytoid cells[a]
1 h	24.0 ± 5.1 (%)	5.1 ± 0.6 (%)	6.7 ± 2.6 (%)	0.44 ± 0.2
18 h	17.4 ± 3.1	3.8 ± 1.7	5.0 ± 1.6	0.34 ± 0.05
3 d	13.6 ± 1.6	2.5 ± 0.0	2.9 ± 0.6	0.47 ± 0.3
5 d	14.0 ± 3.7	2.3 ± 0.8	1.7 ± 0.4	0.24 ± 0.1
10 d	14.2 ± 0.8	3.1 ± 0.5	4.3 ± 2.7	0.48 ± 0.03
14 d	14.6 ± 3.0	3.0 ± 0.3	5.8 ± 2.7	0.43 ± 0.2
20 d	17.6 ± 5.4	4.0 ± 0.7	13.7 ± 2.1	0.7 ± 0.2
30 d	14.7 ± 3.4	2.9 ± 1.1	12.0 ± 2.6	1.1 ± 0.7
45 d	21.8 ± 5.7	5.7 ± 1.7	14.6 ± 2.4	1.9 ± 0.3
60 d	27.8 ± 5.8	9.3 ± 1.8	22.0 ± 3.4	3.4 ± 1.5

[a] Mean ± standard error

5.3 The Proliferative Pattern of Hematopoietic and Stromal Cells in the Red Pulp of Differently Aged Rats

In the following paragraphs the autoradiographic findings (^{3}H-indices) for hematopoietic and stromal cells are described. The mitotic index and the ratio ^{3}H-index: mitotic index, which serve as further proliferation parameters, are also entered on the diagrams.

Large Basophilic Blasts

The mean values of the ^{3}H-indices of these cells fluctuate gently between 60% and 100% without showing distinct minima or maxima. From the 6th–8th week there is a tendency toward a labeling index of 100%. The mitotic index drops from 9.6% immediately after birth to 0 after 60 days. The percentage of mitoses remains on the same low level from the 10th day to the 6th week.

The ratio ^{3}H-index: mitotic index increases until the 10th day and remains at this level until the 45th day (Fig. 23).

Small Basophilic Blasts

Showing a course similar to the large basophilic blasts, the ^{3}H-indices of the small basophilic blasts remain between 60% and 70%. After birth the mitotic index rises slightly to its maximum between the 3rd and 5th day. Then a marked drop sets in, and after 60 days mitoses can no longer be observed.

The ratio labeling:mitotic index remains fairly constant until the 45th day, indicating that in this cell population DNA synthesis is always succeeded by mitosis (Fig. 24).

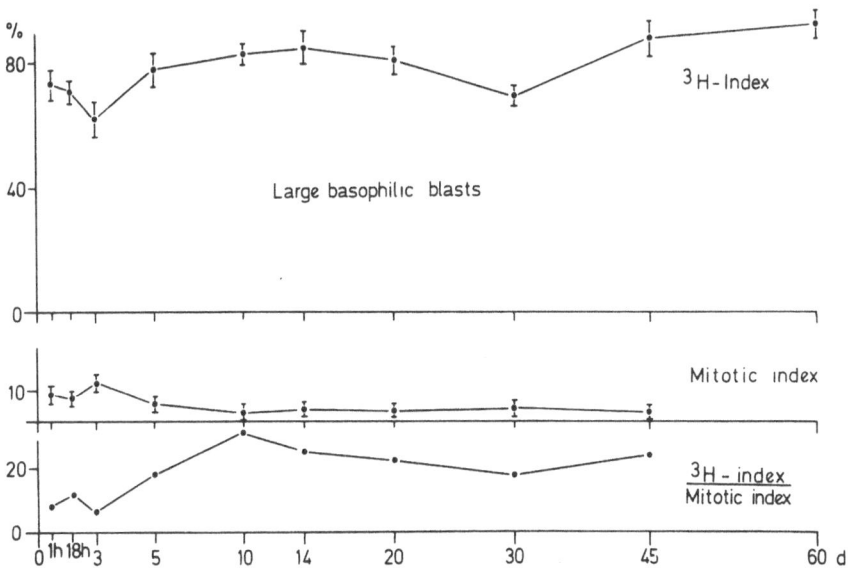

Fig. 23. Large basophilic blasts in the red pulp of growing rats: labeling and mitotic index and ratio labeling:mitotic index

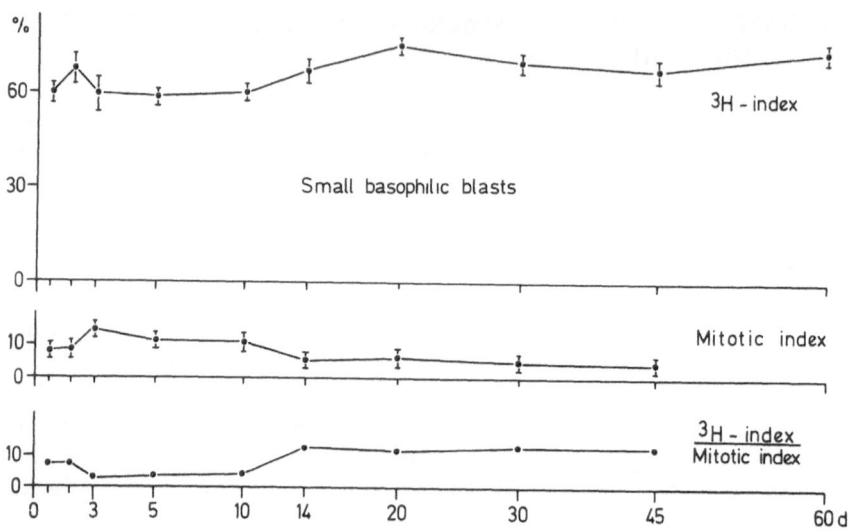

Fig. 24. Small basophilic blasts in the red pulp of growing rats: labeling and mitotic index and ratio labeling:mitotic index

Erythropoiesis

Figure 25 and 26 outline the course of the ^{3}H-thymidine labeling indices of total erythropoiesis and the basophilic and polychromatic erythroblasts, respectively. After an increase until the 5th day, the radioactive labeling remains around 30%–35% until the 45th day. Only toward the end of the experimental time (60th day) is a significant drop apparent.

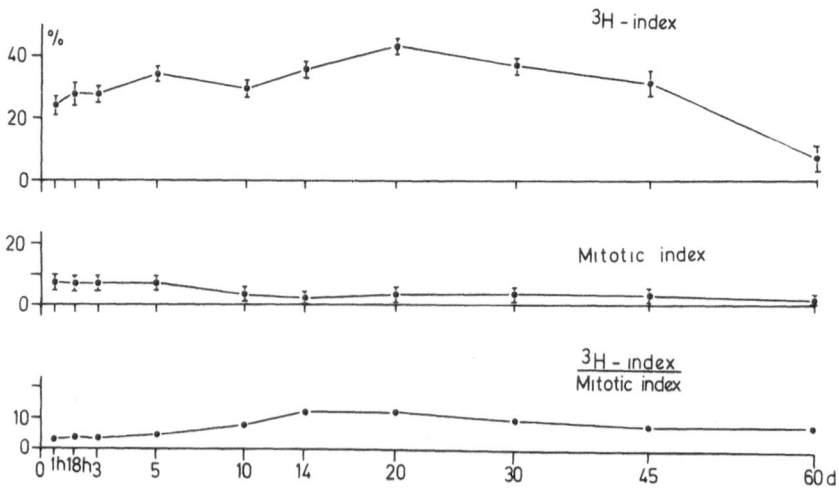

Fig. 25. Total erythropoiesis in the red pulp of growing rats: labeling and mitotic index and ratio labeling:mitotic index

28

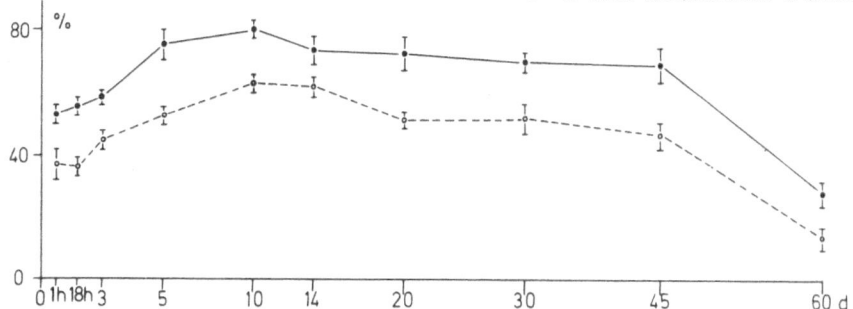

Fig. 26. Labeling index of basophilic and polychromatic erythroblasts in the red pulp of growing rats

The mitotic index, which is 8.6% at the beginning, diminishes only after the 5th day; after 8 weeks it is as low as 1.2%. Compared to the rise in the ^{3}H-labeling index, the drop is more marked between the 10th and 20th day. Thus there is a significant rise in the ratio labeling: mitotic index at this time to a level which is maintained until the 6th week.

A similar labeling pattern of basophilic and polychromatic erythroblasts can be expected. One hour after birth 54.1% of the basophilic and 37% of the polychromatic erythroblasts are labeled with ^{3}H-thymidine. The curves pursue a parallel course to maximal values after 10 days (80% and 64,7% respectively). Then a very gradual drop sets in until significantly lower values (29.1% and 14.1% respectively) are reached by the 60th day.

Under the assumption that every DNA synthesis is succeeded by a mitosis, the ratio labeling:mitotic index equals the ratio $t_s:t_m$. In hematopoietic cells t_m remains fairly constant under various conditions (Killman et al., 1964); in rats it is around 30 min according to investigations of Lord (1970), Monette et al. (1968a) and Roylance (1968). Thus the duration of the S-phase and the cell cycle in erythropoietic precursor cells can be estimated from the counted ^{3}H-TdR and mitotic indices. Under exponential growth conditions, the mitotic index is smaller by the factor ln 2 than under steady state conditions (Maurer and Schultze, 1968), and is then given by the equation $MI = 0.693 \times t_m/t_c$.

The results are shown in Table 7. Both t_s and t_c increase with age. The time intervals are shortest 1 h after birth; by the 5th day they become slightly longer. From the 5th to the 10th day the duration of both the S-phase and cell cycle have doubled and a significant prolongation occurs. The generation time after the 60th day is long, i.e., 28.9 h.

Myelopoiesis

The labeling and mitotic index were only determined on immature myelopoietic cells, not including leukocytes. Then mean values of the ^{3}H-indices remain fairly constant between 20% and 30% (Fig. 27). The mitotic index, which is 3.5% after birth, rises

Table 7. Average duration of S-phase and cell cycle for all dividing erythroblasts (exponential growth pattern) in the red pulp of growing rats. (Duration of mitosis, 30 min)

Age	Proliferating erythropoiesis		t_s (h)	t_c (h)
	^{3}H-indices (%)	Mitotic indices (%)		
1 h	45.6	8.6	1.8	4.0
18 h	46.5	7.9	2.0	4.4
3 d	52.4	7.6	2.4	4.6
5 d	65.0	7.2	3.1	4.8
10 d	72.4	4.1	6.2	8.6
14 d	68.3	2.7	8.7	12.8
20 d	63.2	3.0	7.3	11.6
30 d	62.3	2.9	7.4	11.9
45 d	59.1	2.9	7.0	11.9
60 d	21.6	1.2	6.9	28.9

between 18 and 72 h postnatally (10.6%). The percentage then drops again and remains fairly unchanged until the end of the experimental time. The ratio ^{3}H-index: mitotic index increases slightly after the 5th day.

As described for erythropoiesis, the duration of the S-phase and generation time can also be estimated for granulopoietic precursors in the spleen (Table 8).

According to Rondanelli et al. (1967), Killmann et al. (1964), Boll (1978), and Lord (1965a), the mean value of motosis was taken as 50 min. Whereas S-phase and cell cycle are shortened within the first 3 days, they lengthen to a different extent after the 5th day, reaching maximal values on the 10th and the 60th day, respectively.

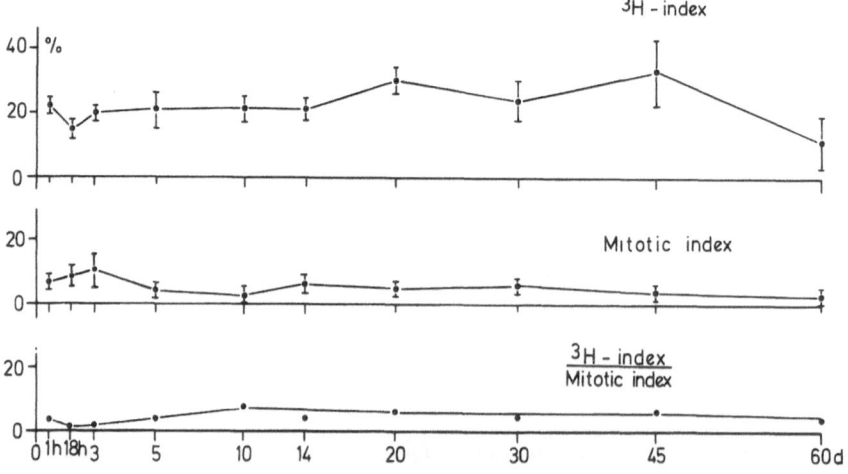

Fig. 27. Total granulopoiesis in the red pulp of growing rats: labeling and mitotic index and ratio labeling:mitotic index

Table 8. Average duration of S-phase and cell cycle for all dividing granulopoietic cells in the red pulp of growing rats. (Duration of mitosis, 50 min)

Age	Proliferating granulopoiesis			
	³H-indices (%)	Mitotic indices (%)	t_s	t_c (h)
1 h	47.5	6.6	4.0	8.4
18 h	41.6	8.2	2.9	7.0
3 d	39.6	10.6	2.1	5.4
5 d	31.0	4.9	3.6	11.7
10 d	37.9	3.0	7.4	19.6
14 d	39.8	6.5	3.5	8.9
20 d	46.5	5.0	5.3	11.5
30 d	36.2	6.4	3.2	8.9
45 d	40.5	4.5	5.2	12.8
60 d	38.5	3.1	7.2	18.6

Reticulum Cells

After 1 h 11.6% of nonphagocytizing reticulum cells are labeled with ³H-TdR. The index remains grossly unchanged until the 10th day. It drops by the factor 2 by the 14th day, and by the factor 3 by the 60th day (3.5%, Fig. 28).

Initially 5.5% of labeled macrophages (phagocytizing reticulum cells) can be found in the spleen. After 18 h, however, the local proliferative activity has diminished to 2.3%; thereafter it ceases completely.

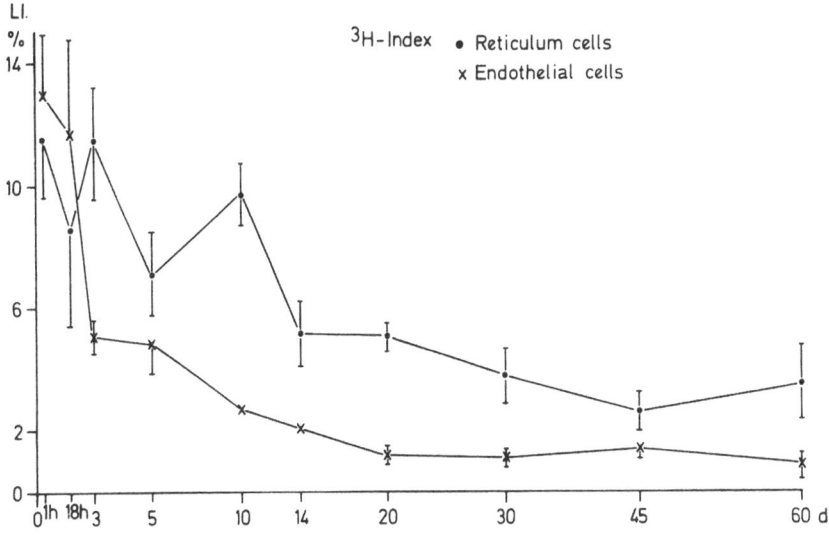

Fig. 28. Labeling index of nonphagocytizing reticulum cells and endothelial cells in the red pulp of growing rats

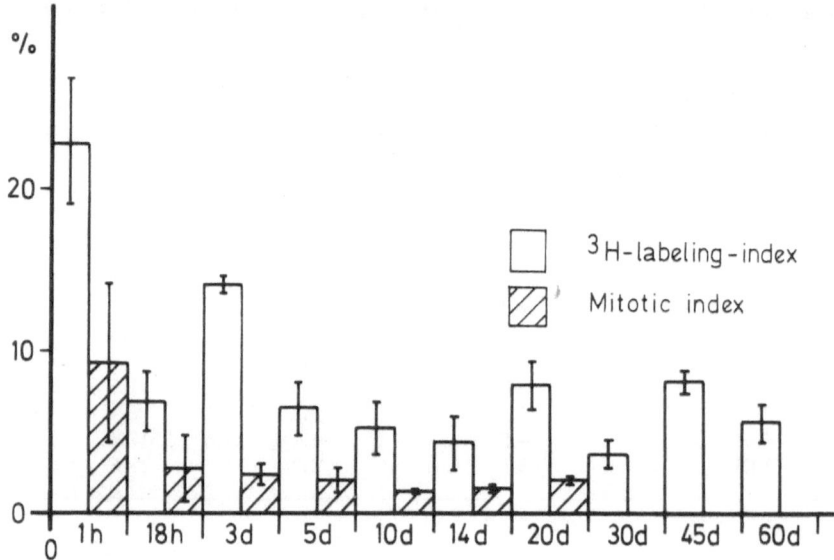

Fig. 29. Labeling and mitotic indices of megakaryocytes in the red pulp of growing rats

Endothelial Cells

The maximal labeling index (11.2% ± 2.8%) is observed 1 h after birth (see Fig. 28). The curve descends rapidly until 36 h (factor 2), then gradually to 0.9% ± 0.4% on the 60th day.

Megakaryocytes

According to Fig. 29, the highest ^{3}H-index is found immediately after birth. A marked decrease has occurred after 18 h to a level which remains rather steady up until 60 days. The course of the mitotic indices is similar, however, and already by the 20th day mitoses can no longer be observed.

5.4 Cell Kinetics of the Red Pulp in Newborn Rats

Table 9 gives the ^{3}H-labeling index and relative amount of different cell populations in the red pulp of newborn Wistar rats.

Basophilic Blasts

One hour after birth the large basophilic blasts amount to 5.5% and the small ones to 10.7%. The initial labeling indices are very close, i.e., 74.9% and 61.2% respectively. Running parallel, the curves reach a maximum of nearly 100% (94%–98%) on the 3rd day, then descend to 0% between the 7th and 10th day (Fig. 30).

Table 9. Percentage distribution and ^{3}H-TdR indices of different cells in the red pulp of neonatal (1-h-old) rats (n = 5)

Cell population	Percentage[a]	^{3}H-TdR index (%)[a]
Large basophilic blasts	5.5 ± 1.9	74.9 ± 4.0
Small basophilic blasts	10.7 ± 2.3	61.2 ± 3.7
Basophilic erythroblasts	7.6 ± 1.0	54.1 ± 1.4
Polychromatic erythroblasts	13.3 ± 1.5	37.0 ± 4.2
Orthochromatic erythroblasts	11.6 ± 1.5	–
Lymphocytes	5.5 ± 0.9	–
Plasma cells	–	–
Reticulum cells without phagocytosis	24.0 ± 5.1	11.6 ± 0.8
Reticulum cells with phagocytosis	5.1 ± 0.6	5.5 ± 1.9
Monocytoid cells	0.4 ± 0.2	–
Endothelial cells	6.7 ± 2.6	11.2 ± 2.8
Granulocytes	2.8 ± 0.6	–
Metamyelocytes	2.3 ± 0.5	–
Proliferating granulopoiesis	4.0 ± 1.1	47.5 ± 2.2
Megakaryocytes	0.5 ± 0.2	23.5 ± 9.1
Percentage of all radioactively labeled cells		21.8 ± 2.0

[a] Mean ± standard error

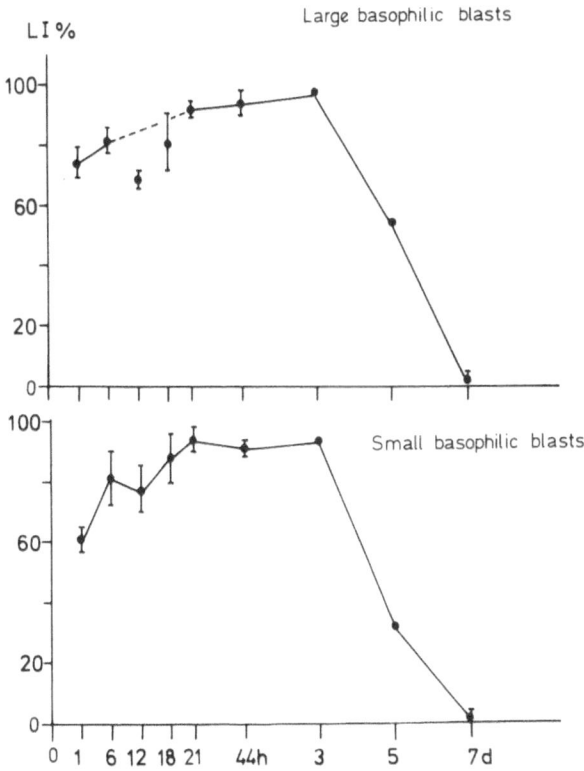

Fig. 30. Labeling indices of large and small basophilic blasts after pulse-labeling with ^{3}H-thymidine

Erythropoiesis

Erythropoietic cells account for 32.5% of the total number of cells in the red pulp of newborn rats. The ^{3}H-labeling index of all nucleus-containing erythropoietic cells, i.e., including orthochromatic erythroblasts, amounts to 23%.

Basophilic Erythroblasts. At the time of birth 7.6% of all cells in the red pulp are basophilic erythroblasts; 54% are in the DNA synthesis phase at that time.

Polychromatic Erythroblasts. The percentage of these cells is 13.3%, the ^{3}H-TdR index 37%.

Orthochromatic Erythroblasts. 1 h postnatally 11.6% of these cells are found in the spleen. Since 1 h after the injection of ^{3}H-TdR nuclear silver grains are not demonstrable, these cells must have lost their ability to synthesize DNA.

Free and already phagocytized erythroblastic nuclei come up to 12.5%, as related to all nucleus-containing erythroblasts (free erythroblastic nuclei 7.8%; phagocytized erythroblastic nuclei 4.7%).

In Figs. 31 and 32 the ^{3}H-indices of different erythroblasts were plotted against various time intervals after ^{3}H-TdR injection. The labeling index of the basophilic erythroblasts rises quickly, reaching its maximum after 21 h. This is followed by a gradual descend until the 3rd day and a steeper one until the 7th day.

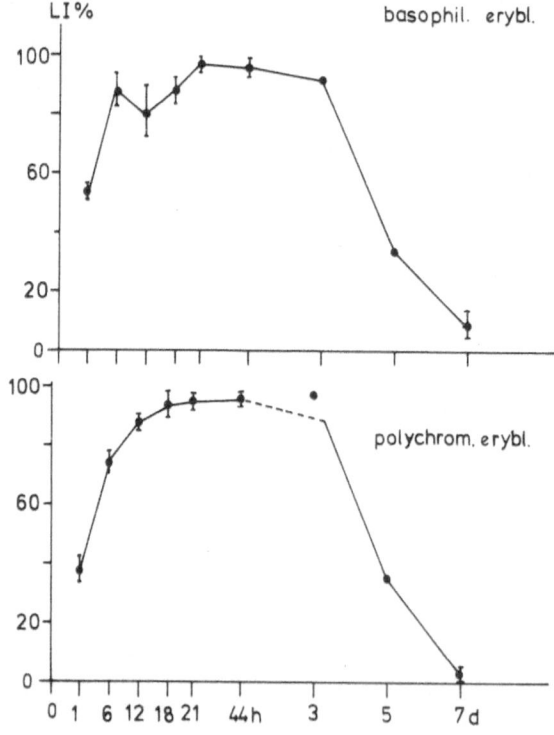

Fig. 31. Labeling indices of basophilic and polychromatic erythroblasts after pulse-labeling with ^{3}H-thymidine

34

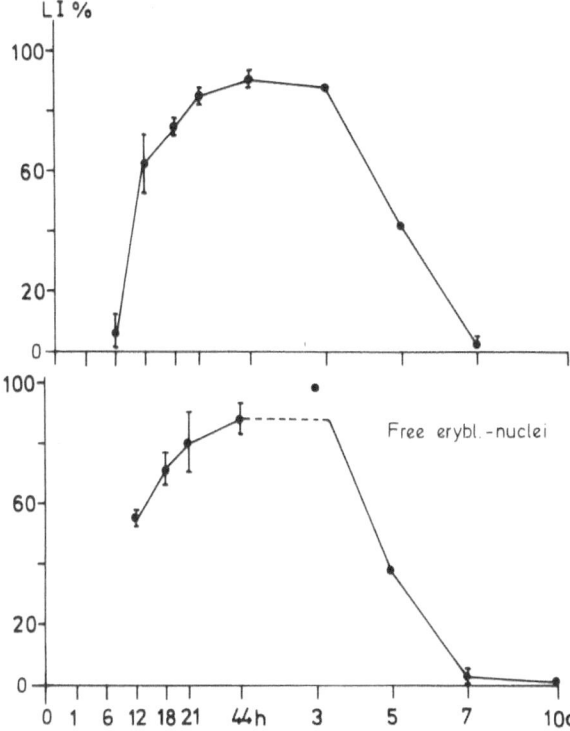

Fig. 32. Labeling indices of orthochromatic erythroblasts and expelled erythroblast nuclei after pulse-labeling with ^{3}H-thymidine

The labeling indices of the polychromatic erythroblasts pursue a parallel course with an almost equal maximal value.

The first radioactively labeled orthochromatic erythroblasts appear after 6 h. Their number increases rapidly, reaching the maximum of 90.5% after 44 h. This is followed by a gradual descent until the 3rd day and a steep descent to 0% by the 7th day.

The curve of labeled expelled erythroblast nuclei has a similar course. The first labeled nuclei can be observed 6 h later, i.e., after 12 h. They amount to 50% initially and reach their maximum after 3 days. Until the 7th day a rapid decrease occurs.

Table 10. Cell fractions and ^{3}H-TdR labeling indices of granulocytic precursor cells in the red pulp of 1-h-old rats

	Promyelocytes	Myelocytes	Metamyelocytes
Fraction of total	0.23	0.49	0.28
labeling index (LI)	0.52	0.43	–

From the above-mentioned parameters the hourly progression rate can be calculated: for orthochromatic erythroblasts it is 2.1%, and for free erythroblast nuclei around 1.6%—2.0%.

The ratio between the erythroblasts is the following: basophilic erythroblasts 19%, polychromatic erythroblasts 33%, orthochromatic erythroblasts 40%, and proerythroblasts (hypothetical) 8%.

Granulocytopoiesis

Proliferating Myelopoiesis. The amount of immature myelopoietic cells in the spleen of 1-h-old rats is 4.0%, and the labeling index 47.5%. The latter increases to a maximum of 80%—88% between 44 and 72 h. Then there is a steep drop to values below 10% after 7 days (Fig. 33).

The percentages of the different immature granulopoietic cells given in Table 10 are mean values of five animals.

Non-proliferating Myelopoiesis. There are 2.3% of metamyelocytes and 2,8% of granulocytes in the neonatal spleen.

Radioactively labeled metamyelocytes are present after 12 h, labeled granulocytes after 18 h. The labeling indices of both cell types exhibit a parallel course with the

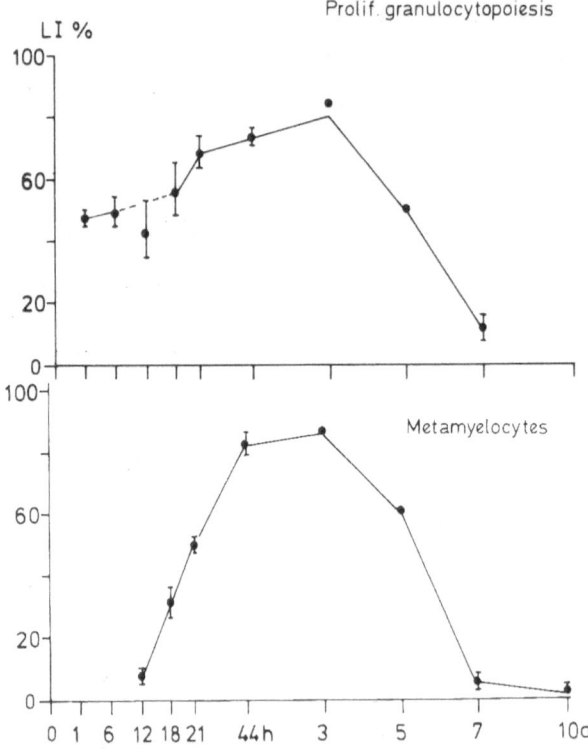

Fig. 33. Labeling indices of proliferating granulopoietic cells (promyelocytes, myelocytes) and metamyelocytes after pulse-labeling with ^{3}H-thymidine

maximum for granulocytes (65%) staying far below the one for immature myelopoie-tic cells and metamyelocytes (Figs. 33, 34). From the increase in the labeling indices the progression rate can be calculated: 2.2% metamyelocytes/hour, 2.1% granulocy-tes/hour.

Lymphocytes

The red pulp of the neonatal rat spleen contains 5.5% ± 0.9% of lymphocytes (Table 9). Labeled lymphocytes appear after 12 h. After 18 h there is a first peak of immigrating lymphocytes, and after 3 days another one (67%, Fig. 34). From the first ascent the hourly inflow rate is calculated as 4.3%, from the second one as 0.9% (21.6%/day).

Plasma Cells

In the neonatal spleen plasma cells could not be found.

Reticulum Cells

Non-phagocytizing Reticulum Cells. 24.0% of the cells in the red pulp of newborn rats are nonphagocytizing reticulum cells. Initially the labeling index is 11.6%, but af-ter 44 h it has significantly increased. It then assumes a plateau-like course and drops

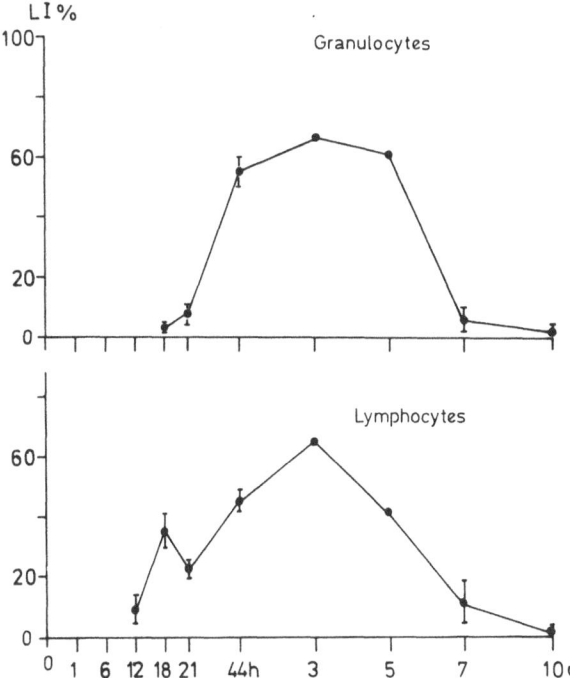

Fig. 34. Labeling indices of granulocytes and lymphocytes after pulse-labeling with ^{3}H-thymidine

37

markedly until the 7th day. After 4 weeks 2% of the cells are still radioactively labeled (Fig. 35).

Phagocytizing Reticulum Cells. 1 h after birth they amount to 5.1%. The labeling index, which is 5.5% initially, rises by the factor 7 until the 12th hour. Until the 3rd day it fluctuates around 24%–26%. After a gradual decrease, labeled macrophages are observed only occasionally 4 weeks after birth (around 2%) (Fig. 35).

Monocytoid Cells

Because of their infrequent occurrence in the spleen of newborn and adult rats (0.4%, Table 9), the ^{3}H-indices could not be determined.

Endothelial Cells

The ^{3}H-indices rise from 11.2% initially to maximal values on the 5th day. After a marked drop, the labeling index remains around 1.5% from the 12th to the 28th day (Fig. 36).

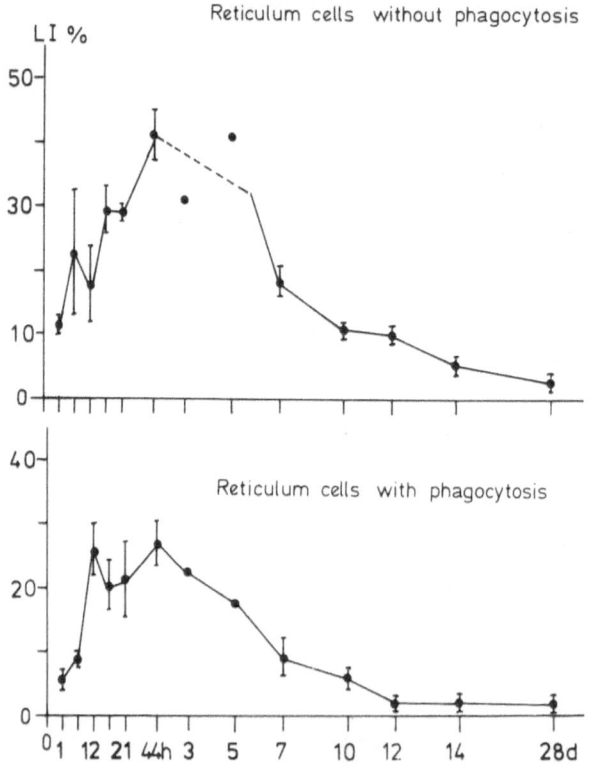

Fig. 35. Labeling indices of nonphagocytizing and phagocytizing reticulum cells after pulse-labeling with ^{3}H-thymidine

38

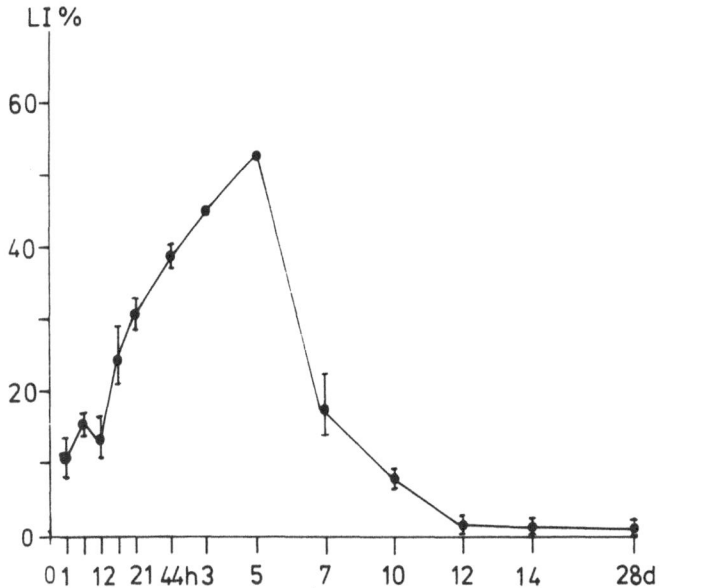

Fig. 36. Labeling index of endothelial cells after pulse-labeling with ^{3}H-thymidine

Megakaryocytes

Of these cells which amount to 0.5%, 23.5% have initially incorporated ^{3}H-thymidine. The mitotic index is 9%. The highest value of 78.9% is reached after 44 h, being followed by a rapid, then gradual drop until the 7th day (Fig. 37).

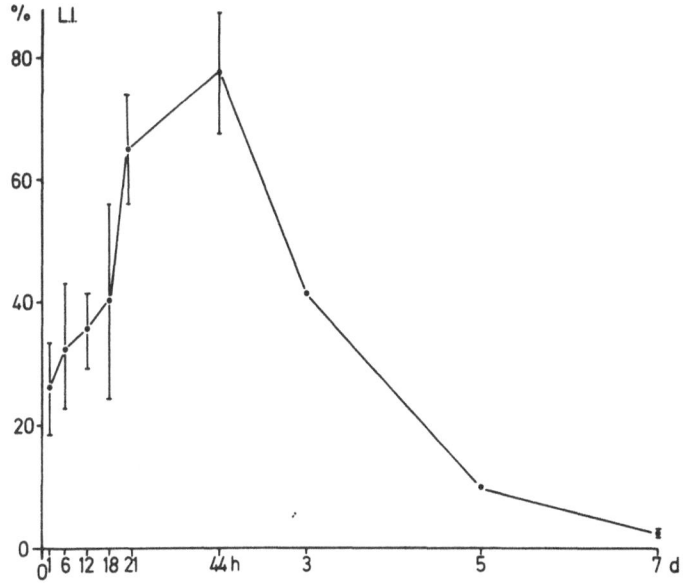

Fig. 37. Labeling index of megakaryocytes after pulse-labeling with ^{3}H-thymidine

6 Discussion

6.1 Classification and Frequency of Hematopoietic and Stromal Cells in the Red Pulp of Differently Aged Rats

6.1.1 Basophilic Blasts

They are large (more than 10 μm) or small (less than 10 μm), otherwise uniform immature cells which are frequent in the postnatal spleen between 18 and 120 h. They have been described extensively by Lennert (1961) and Mori and Lennert (1969).

A part of the basophilic blasts, notably the small ones, has to be subordinated to the group of proerythroblasts and myeloblasts (Yamashita and Helpap, 1974b). In contrast to studies on smears and electron microscopic studies, these earliest cells of erythropoiesis and granulocytopoiesis, respectively, cannot be identified in the semithin section. In the histologic section they may be distinguished only because of different locations. Proerythroblasts group to form small aggregations, sometimes together with more mature erythroblasts; myeloblasts are found isolated in the peritrabecular area or close to myelopoietic cell aggregations in the subcapsular region (Lennert, 1952; Orlic et al., 1968).

By analogy to extensive investigations on the bone marrow, it can be concluded that the remaining primitive basophilic blasts in the red pulp are stem cells, i.e., precursor cells which are already determined for a specific blood cell system (Rosse, 1976; Rosse and Yoffey, 1967; Rosse and Trotter, 1974).

Lennert (1952) supposed that the large hemocytoblast, in contrast to the small one, could be a still more immature (erythropoietic?) precursor. Recent investigations on the stem cell problem have disproved this assumption. Today there is no doubt that the most immature hematopoietic cell – the pluripotent stem cell which all blood cell systems including T- and B-lymphocytes have in common – resembles in the non-active stage a small lymphocyte. This cell is able to transform into the proliferating type of the transitional cells and into "blasts" (Fliedner, 1975; Rosse, 1976; Dicke et al., 1973; Hoelzer et al., 1975). Hence these nonactive pluripotent stem cells should be found among the small lymphocytes in the red pulp.

Electronoptical investigations by Yamashita and Helpap (1974c), Orlic et al. (1965), Rifkind et al. (1969), and Zamboni (1965) have indicated that part of the large basophilic blasts serve as precursors of the megakaryocytes.

In the literature no comparable details are given regarding the frequency of the basophilic blasts in the postnatally developing rat spleen. According to Fig. 14, the frequency becomes less with increasing age; single cells are still present at the end of the 2nd month.

Since the nonactive pluripotent hematopoietic stem cell resembles a small lymphocyte, it could not be identified in the present study. Besides, under normal conditions the amount of these stem cells in the spleen is very low. In adult mice the ratio of stem cells to the total number of cells is 1:100 000 (McCulloch and Till, 1963), while in 2-month-old rats it is 5:1 000 000 (Vacek et al. 1976).

The number of erythroblasts in the spleen increases until the 10th day. After the 30th day it diminishes gradually, and after 8 weeks 4% are still left. The marked rise until the 10th day agrees with the splenograms of newborn rats by Lucarelli et al. (1966).

However, the amount of 31.1% in 4-week-old animals found in the present study, is about twofold. This difference is probably due to a different methodical procedure. The above-mentioned splenogram of Lucarelli et al. (1966) was carried out on spleen smears considering white and red pulp cells equally. As the percentage of "lymphoreticular cells" increases steadily at this time corresponding to the fully developing white pulp, the relative amount of erythropoietic precursors consequently decreases. The actual number of lymphocytes in the red pulp on the 30th day is much lower, being 12.6% in the semithin section.

According to the methods of cell system physiology, erythropoiesis is presently considered as a series of successive cell pools (Fliedner and Calvo, 1969; Fliedner et al., 1969). There is a stem cell, a proliferation, a maturation, and a functional pool. Weicker (1954a, b) proved in extensive investigations that there is a certain quantitative relationship between the erythroblasts. Generally, the size of the cells at each of the different stages decreases as a function of maturation while their number double.

The ratio proerythroblast:macroblast:basophilic erythroblast:polychromatic erythroblast:orthochromatic erythroblast equals 1:2:4:8:16. Thus under normal conditions 16 erythrocytes originate from one proerythroblast (Bessis, 1973; Fliedner et al., 1969). Haas et al. (1970) found the same pattern in liver erythropoiesis of newborn rats, which ceases soon after birth.

This principle of doubling cell amounts cannot be confirmed if the relative amounts of the erythroblasts in the spleen are considered (Fig. 16). With respect to the polychromatic erythroblasts, the orthochromatic erythroblasts are too low in number, being only about half of the expected amount.

The phenomenon of ineffective erythropoiesis, which alters the relationship between the erythropoietic stages considerably, could explain this finding (Messner et al., 1969; Fliedner et al., 1969). Under normal conditions it is 5%–10% in man; under pathologic conditions such as hemolytic anemia, thalassemia major, pernicious anemia, and sideroblastic anemia it plays a more important role (Bessis, 1973; Fliedner et al., 1969; Wickramasinghe et al., 1967, 1976); this is also so in extramedullary hematopoiesis (Haurani and Tocantins, 1961; Neumann et al., 1976; Wekel et al., 1970). Ineffective erythropoiesis is characterized by the decay of erythroblasts on various maturation stages or maturation to the next stage without division (Fliedner et al., 1969). The perishing erythroblasts which have been eliminated from the proliferating pool are phagocytized by reticulum cells. This should cause an increased number of free and phagocytized erythroblast nuclei (Bessis, 1973). Also, the number of polychromatic erythroblasts should become less. Investigations on the rat spleen do not confirm this assumption. The percentage of polychromatic erythroblasts remains constantly high, and the number of expelled erythroblast nuclei decreases continuously after birth to 1.6%–3% from the 10th day onward.

Another explanation appears more likely: obviously a great many of the polychromatic erythroblasts in the rat spleen do not divide further, but transform directly into orthochromatic erythroblasts, thus leading to equal percentages of poly- and orthochromatic erythroblasts. This "shift to the left" indicates a strong stimulation of erythropoiesis in the spleen. Under omission of further cell divisions, polychromatic erythroblasts are allowed to enter the maturation compartment prematurely, as has been observed in human erythropoiesis under conditions of stress like hemorrhage or leukemia (Hoelzer et al., 1972; Killmann, 1970; Doermer, 1973). Evaluating myelograms of newborn human infants, Ruhrmann (1966) assumed that some polychromatic erythroblasts do not pass the stage of the youngest normoblast, but reach the blood as reticulocytes after elimination of the nucleus. Warninghoff and Hausmann (1955) obtained the same finding in the hematopoiesis of human embryos. Under these circumstances short-lived reticulocytes and erythrocytes with a much bigger volume are produced, i.e., macrocytes (Stohlman et al., 1964; Killmann, 1970). In rats stimulated erythropoiesis can be assumed to be due to postnatal anemia, with the maximum around the 15th–20th day (Lucarelli et al., 1968a).

It is an interesting observation that the macrocytosis disappears after the 60th day, i.e., at a time when the erythropoiesis in the spleen has nearly completely ceased.

In any case the spleen can be regarded as a production place for macrocytes which are characteristic of the postnatal stage of rats and premature human infants.

6.1.3 Myelopoiesis

The myelopoietic activity in the rat spleen is only a third of the erythropoietic one; this finding applies also to the extramedullary hematopoiesis in the human spleen (Sjögren, 1978; Söderström et al., 1975). Curry and Trentin (1967) proved by means of the spleen colony technique (Till and McCulloch, 1961) that 60% of the spleen colonies consisted of erythropoietic cells, and only 20% of the neutrophilic myelopoietic cells. They concluded that – in contrast to the stroma of the bone marrow – the spleen stroma allows mainly the development of erythroid cells. Taking this into consideration, the dominance of the erythropoiesis over the granulopoiesis in the rat spleen was to be expected. Myelopoietic proliferating cell aggregations have a characteristic subcapsular and peritrabecular location. In rare instances they can be found inside atrophic lymph follicles (Curry and Trentin, 1967).

There is only a small amount of information in the pertinent literature on the spleen granulocytopoiesis in growing rats. Kindred (1942) observed in 15- and 20-day-old rats about 1% neutrophilic granulocytes, 1.2%–3.2% eosinophilic granulocytes and 2.9%–6% myelocytes. According to Lucarelli et al. (1966), the percentage of myelopoietic cells in the spleen remains fairly constant till the 32nd day. In the present study 6.3% immature myelopoietic cells and 2.8% leukocytes were found 1 h after birth. The fluctuations during the experimental time and the differences from findings by other investigators are probably due to the individual reaction of the animals toward intercurrent infections, since they were not kept under sterile conditions. Besides, it has to be considered that some of the leukocytes originate from the bone marrow and are transported via the blood stream into the spleen.

Whereas the immature proliferating myelopoietic cells decrease in number in the spleen after the 5th day, the leukocyte increase corresponds to their rising numbers in the peripheral blood (Plum, 1943).

42

6.1.4 Lymphocytes

Splenograms of rats described in the literature considered white and red pulp equally, thereby yielding much higher values for lymphocytes than the present selective study. Therefore a comparison is not possible. Riman et al. (1958) observed 60% in 1-day-old rats, which increased to 95% in adult ones. Lucarelli et al. (1966) and Richter (1953) found similar amounts in dogs and rabbits. Considering the red pulp selectively, 8% were found in younger animals and 16%–20% in older ones; the latter comes close to the value for the bone marrow (Ramsell and Yoffey, 1961; Reincke, 1962).

As pointed out before, a significant increase in lymphocytes occurs postnatally, with steady concentrations after the 14th day. Obviously the amount of recirculating lymphocytes leaving the white pulp does not change anymore at this time and already corresponds to the situation in adults.
Also, the B and T cell regions in the white pulp are fully functioning at this time (Veerman, 1975).

Both T- and B-lymphocytes pass the red pulp with T-lymphocytes migrating much faster (5–6 h) than B-lymphocytes (days). Therefore it can be presumed that predominantly B-lymphocytes are present in the red pulp. Investigations by Parrot and de Sousa (1971) support this assumption: they demonstrated that bone marrow (B) lymphocytes which were labeled with ^{3}H-adenosine stayed predominantly in the red pulp. Furthermore, comparative studies with ^{3}H-TdR and ^{3}H-CdR labeled lymphocytes confirmed this finding (Helpap and Dachselt, 1978).

While the lymphocytes in the red pulp of the rat spleen remain at the level of the adult spleen after the 14th day (Fig. 20), erythropoiesis decreases (Fig. 15). This parallels the conditions in the human infant where the bone marrow exhibits a strong infiltration with lymphocytes at a time when the erythropoiesis diminishes (Ruhrmann, 1966). The same applies to chronic hypoplastic anemias in childhood where the lymphocytes drop in number after corticosteroid medication. Finally, it is well known that thymus extracts and also thymus tumors may suppress the erythropoiesis and lead to erythroblastopenias ("pure red-cell aplasia"; Queißer, 1978). Corsi and Giusti (1967), Trainin and Resnitzky (1969), Resnitzky and Trainin (1971) and Miller et al. (1965) could actually observe an increase in undifferentiated blasts and early erythroblast stages after neonatal thymectomy. However, a stop in the maturation process and suppression of the proliferation of pluripotent stem cells apparently existed at the same time.

6.1.5 Reticulum Cells

The data on the reticulum cells in the red pulp disagree considerably, primarily due to different techniques (paraffin sections, smears) and different definitions. We considered fixed stromal cells, which form a meshwork in the spleen (Lennert and Müller-Hermelink, 1975), to be reticulum cells. Functional differences allow a distinction between two types: nonphagocytizing and phagocytizing (histiocytic) reticulum cells (Mori and Lennert, 1967; Müller-Hermelink and Lennert, 1978).

Nonphagocytizing Reticulum Cells. There is a wide range in numbers given by different authors: Riman et al. (1958) found 0.47% and 0.25% reticulum cells in splenograms of newborn and adult rats, respectively. Kindred (1940) observed 11.6% in 15-day-old rats, 8.8% in 20-day-old rats.

Taking the topography of the spleen into account and referring to unit areas, the concentration of the nonphagocytizing reticulum cells appeared fairly constant throughout different ages (Fig. 22). Obviously the amount is kept on a constant level by precursor cells which immigrate according to the enlargement of the spleen.

Phagocytizing Reticulum Cells. The concentration of macrophages in the red pulp remains rather constant till the 45th day and increases significantly only toward the end of the experimental time (60th day; Fig. 22). Thus under normal conditions the concentration of these cells remains stable in spite of the growing organ. Phagocytic activity and capacity remain unchanged and are not submitted to a stronger stimulation. Culbertson (1939) did not observe any differences in the phagocytic activity of spleens from 6–60-day-old rats. Probably the phagocytic system in the rat is only slowly activated with increasing age and fully developed only after the 60th day (McFadden, 1966). This could explain the increase in the number of macrophages at this time.

6.1.6 Endothelial Cells

As fixed and not easily detachable cells they occur only rarely in smears and imprint preparations and consequently there are no data given in the literature. The concentration remains unchanged until the 10th day, increases threefold by the 20th day and remains at this level. Investigations by Stutte (1974) have demonstrated that an increased intrasplenic blood volume, e.g., in congestive heart failure, also causes an increase in the sinus endothelial cells. The increase in the numbers of endothelial cells between the 10th and 20th day happens at a time when the red pulp has morphologically matured and indicates that an intensification of the circulation no longer occurs. The circulatory system has probably been reorganized during this time interval, as it happens in a similar manner in the bone marrow during the trimenon reduction of the newborn (Ruhrmann, 1966; Riegel and Ruhrmann, 1964).

6.1.7 Monocytoid Cells

They are only occasionally found in the red pulp, but the number rises by the factor 10 with increasing age. Comparable details are only given by Riman et al. (1958) who did not find an increasing number with growing age. They observed 0.2% monocytes in the splenograms of newborn rats (0.4% in the present study), 0.12% in 80-day-old animals (in this study 3.4% ± 1.5% in 60-day-old animals). Richter (1953) found 1.5% and 1.7% in young rabbits and dogs respectively. In adult rats the spleen (and lungs) are the preferential sites of monocyte concentration (Whitelaw and Batho, 1972).

6.1.8 Plasma Cells

The present values (0.2%–1.5%) agree with the values in the literature (0.75% ± 0.14%; Tischendorf, 1969).

The present results concur with those of Marien and McFadden (1968), who also determined maximal amounts on the 14th day.

6.2 The Proliferative Activity of Hematopoietic and Stromal Cells in the Red Pulp of Differently Aged Rats

6.2.1 Basophilic Blasts

From the pattern of the labeling and mitotic indices and the cell number the following conclusions can be drawn:

1. The basophilic blasts in the rat spleen remain an intensely proliferating cell population until adulthood.
2. There are differences in the proliferative pattern: Between the 3rd and 5th day large basophilic blasts exhibit a high DNA synthesis rate, but diminishing divisional activity. At this time small basophilic blasts exhibit a constant ratio of DNA synthesis rate to mitotic rate; the latter diminishes only between the 10th and 14th day. This pattern is similar to the splenic erythropoiesis.
3. The number of cells decreases with growing age. Since the multipotent stem cells are chiefly nondividing resting cells (Lajtha, 1975), the constantly high DNA synthesis rate of the basophilic blasts implies that they have already originated from the pool of multipotent stem cells.

   Determined stem cells, which comprise part of the basophilic blasts, are actively proliferating and dividing cells (Fliedner, 1974, 1975, 1976; Rosse, 1976). Under the influence of stimulating humoral factors (erythropoietin, thrombopoietin), these cells supply the systems of the erythro-, granulo-, megakaryo- and lymphopoiesis. Since only half of the cells differentiate specifically, whereas the other half remains in the stem cell pool, the number of cells in the pool is maintained (Fliedner, 1976).

   Though the divisional activity of the large basophilic blasts diminishes (decreasing mitotic index between the 3rd and 5th day), the DNA synthesis rate remains the same. Obviously this maintains the sensitivity of the cell system toward hematopoietic regulating substances, thus ensuring a fast activation of the hematopoiesis under pathologic conditions (hemorrhage, stress, radiation; Hanna, 1967; Lord, 1967; Rosse, 1976). Since this cell population proliferates rapidly even under normal conditions, a higher turnover is only possible with an increased stem cell pool. Without stimulating influences, a certain part of the proliferating stem cells (large basophilic blasts) could theoretically turn into nondividing stem cells in order to reduce the active pool (Rosse, 1976; Haas et al., 1970). Resting and determined stem cells might also be induced by age-dependent alterations in the "hematopoietic-inductive-microenvironment" (Trentin, 1971) to emigrate from the spleen into the bone marrow (?) which offers more favorable conditions for proliferation (Metcalf and Moore, 1971; Silini et al., 1976).

Labeling Pattern. According to Lord (1965a), who examined spleen smears of male August rats, the ^{3}H-TdR index remains on a steady level from the 10th to the 18th day of life reaching 56% after 10 days; then it drops rapidly to 20%–30% between the 20th and 30th day. The present investigation yielded somewhat different results: from the 10th to the 45th day the percentage of radioactively labeled erythroblasts fluctuates around 30%; a significant decrease sets in 14 days later than in Lord's observation. This disagreement is probably due to the separate evaluation of proerythroblasts in the spleen smears, which was – due to technical reasons – not possible in this study. From birth until 7 weeks afterwards proerythroblasts in the spleen exhibit a high labeling index of 70%–90% (Lord, 1965a); in the bone marrow it amounts to 70% (Tarbutt and Blackett, 1968; Roylance, 1968). The same value applies to small basophilic blasts, thereby giving another indication of their possible proerythroblast nature.

Genetic or sex-specific differences between various breeds might also account for differing results; however, not to the same extent as in mice (Yamashita and Helpap, 1974b).

According to the investigations of Lord (1965a), the labeling index of splenic basophilic erythroblasts is 70%–75% between the 10th and 25th day, and is thus in good agreement with the results of the present study. In contrast, however, these high values are still demonstrable after the 45th day (Lord: 50%). The range of the ^{3}H-TdR index of the bone marrow of 6-week-old (adult) rats is between 55% and 96% (Tarbutt and Blackett, 1968; Roylance, 1968; Heininger et al., 1971). The corresponding values for polychromatic erythroblasts in the bone marrow amount to 30%–67% (Tarbutt and Blackett, 1968; Roylance, 1968; Heininger et al., 1971; Hanna, 1967).

The decline in the erythropoiesis with growing age is due to the gradual transition of a proliferating to a resting stem cell pool (Porcellini et al., 1976), as mentioned under "basophilic blasts." Even after 6–8 weeks all the maturation stages can be observed in the spleen, similar to the liver erythropoiesis at an earlier stage. A sudden stop in the proliferation and differentiation of the stem cells in the spleen does not happen; this would lead to a gradual disappearance of immature cells with eventually only mature cells left (Haas et al., 1970).

Cell-kinetic Parameters. The mitotic index of the erythropoiesis in the spleen follows a different pattern from the labeling index. The divisional rate drops from 8.6% 1 h after birth to 2.7% on the 14th day. After 45 days there is a further decrease to 1.2% on the 60th day. This means that during the process of maturation the divisional capacity of the erythroblasts becomes gradually less, i.e., the average generation time is prolonged (Lord, 1965a). A theoretically possible prolongation of the mitotic duration itself cannot be made responsible, since it remains rather stable under various conditions (Killmann et al., 1964). From the increasing or constantly high ^{3}H-TdR index and the simultaneously and parallel decreasing mitotic frequency it has to be concluded that the S-phase changes in duration from birth to the 10th day of life. On the 10th day generation time and S-phase have doubled (Table 7). It is also shown that from the 10th to the 60th day the duration of the S-phase remains within the 6- to 8-h range for t_s, which has been found for dividing hematopoietic and various nonhematopoietic tissues in different animal species (Maurer and Schultze, 1968; Monette et al., 1968). The changes in the length of the cell cycle are due to the variable duration of G_1 and/or G_2 phases. By contrast, during the first 5 days very short S-phases and generation times are also observed, corresponding to those of fetal erythroblasts in the liver of mice and rats (Tarbutt and Cole, 1970; Lucarelli et al., 1968b).

It has been demonstrated that the S-phases and generation times of erythroblasts are variable parameters and shorten after acute hemorrhage or chronic anemias (Hanna et al., 1969). A prolongation occurs with increasing age (Roylance, 1968; Tarbutt and Blackett, 1968; Porcellini et al., 1976) and has also been observed in iron deficiency anemias in man (Dörmer and Lau, 1977, 1978). Porcellini et al. (1976) determined the cell cycle parameters of proeythroblasts and basophilic erythroblasts in the bone marrow of 5-, 10-, and 20-day-old and grown-up rats. They proved that the generation time increased from 7.5 h in young rats to 10 h in adult rats.

According to Lord (1967), the erythroblast cycle in the spleen will last for about 15 h after stimulation like acute bleeding. In anemic rats the spleen may account for much as 20% of the total red cell production (Tarbutt, 1969; Blackett, 1971). Under this stimulation, the stem cells in the spleen are proliferating and producing erythrocytes at the maximum rate. To approve of this value one must consider that under normal conditions the spleen of adult animals shows none or very little erythropoietic activity with cell cycles of about 29 h. The same result for the bone marrow of adult rats (32–58 h) was obtained by Lord (1965b).

A prolonged cell cycle combined with an unaltered cell maturation, means a decrease in the cell division and production (Lord, 1965a, 1967; Hanna et al., 1969). In addition, the cellular inflow from the stem cell compartment is gradually reduced. Accordingly, the erythropoietic cell production in the spleen decreases continuously provided there is no additional stimulation. As the mitotic indices of the basophilic blasts and the total erythropoiesis indicate, this should happen between the 3rd and 10th day of life, at a time when the bone marrow hematopoiesis is strongly activated (Lucarelli et al., 1964).

6.2.3 Myelopoiesis

Structure, function, and regulation of the granulopoiesis under normal and pathologic conditions have been described by Cronkite and Vincent (1969), Cronkite and Fliedner (1964), and Fliedner (1974). As in erythropoiesis, they distinguish a stem cell, proliferation, maturation, and functional pool. However, in particular the knowledge about regulation and feedback mechanisms in the myelopoiesis leaves a number of unsolved questions.

Lord (1965a) found 11% ^{3}H-TdR labeled myelopoietic cells in the spleen between the 10th and 18th day, a decrease to 8% on the 25th day and to 4% on the 35th day. Showing the same course, the results for the bone marrow are higher and drop only as far as 15%. The period from birth until the 10th day was not investigated. In contrast to the present investigation, Lord (1965a) calculated the labeling index with regard to the total myelopoiesis including polymorphonuclear granulocytes. This explains the difference to the present results with an average of 20%–30%. Under consideration of the granulocytes, the labeling index is also 8%–13%, but a significant drop occurs only between 6 and 9 weeks after birth.

The mitotic indices of the myelopoietic cells in the spleen present a different behavior: an increase until the 3rd day, then a descent and a constant level from the 5th to the 45th day. The contact of the organism with a nonsterile environment and infectious agents after birth obviously leads to a higher demand for granulocytes (Metcalf and Stevens, 1972). The initially increased mitotic frequency means a shortening of the generation times (G_1 and S-phase; Cronkite and Vincent, 1969). After the 5th day the mitotic frequency is lower again, i.e., the generation times are longer.

The data in Table 8 support the idea of such a regulating mechanism. The duration of 4 h for the DNA synthesis in newborn rats corresponds to the 5 h obtained in the bone marrow of young dogs (Maloney et al., 1963; 1971; Patt and Maloney, 1964). The shortened S-phase (2—3 h) and generation time (5—7 h) during the following 3 days of life exhibit — similar to the stimulated erythropoiesis — an increased postnatal demand for granulopoietic cells. Later there are no significant differences from the values for the bone marrow of rats of the same age (t_s 7.8—8 h, t_c 11.5—12 h; Constable and Blackett, 1972). Lord (1965b) determined cell cycles as long as 22—57 h for myelopoietic precursors (myeloblasts and myelocytes) in the bone marrow of untreated rats.

6.2.4 Reticulum Cells

The proliferation of the nonphagocytizing reticulum cells is most intensive during the first few days after birth, then it drops more and more until the labeling index is only 3.5% 60 days after birth. The ^{3}H-index in the spleen of adult mice was determined as 1.7% (Yamashita and Helpap, 1974b). On the whole, turnover and divisional activity of these cells are very low (Haas et al., 1969a, b; Fliedner et al., 1968a, b).

After 18 h radioactively labeled macrophages can no longer be observed in the spleen; a local proliferation does not take place any more. They are obviously replaced by immigrating cells (monocytes) from the bone marrow.

6.2.5 Endothelial Cells

The radioactive labeling pursues the same course as in reticulum cells: the maximum during the first 18 h is followed by a rapid drop, on the 60th day the index is only 0.9%. Such a low ^{3}H-index was also observed in the sinus endothelial cells of the bone marrow (Tavassoli, 1975). On the whole, only a small percentage of this cell population is in the S-phase (except in the perinatal period); however, the activity can be stimulated (e.g., by injection of urethane; Langdon and Berman, 1975). On the other hand, a low turnover rate was also observed in neonatal rats. In rats, which were completely radioactively labeled at birth, the ^{3}H-index did not drop within the first 2 weeks; only after 10 weeks had it diminished to 40%—50% to remain at this level for a further 5 months (Haas et al., 1969a).

6.2.6 Megakaryocytes

Unlike the proliferating cell systems of the basophilic blasts, erythropoiesis and myelopoiesis, which show postnatally increased or at least constant values of the ^{3}H-indices, the labeling index of megakaryocytes has already dropped by the factor 4 after 18 h, thereafter remaining almost unchanged.

Only the type I megakaryocyte is capable of synthesizing DNA. Since the cells do not divide subsequently, endoreduplication of the chromosomes occurs and polyploid cells are produced (Queißer, 1978). The megakaryocytes, which have lost the capability of dividing, must be preceded by a precursor cell which proliferates under stimulation. In that case, the type I cell increases in size, the maturation time is shortened, and the ^{3}H-index is elevated (Blackett, 1971; Ebbe et al. 1968; Grouls and Helpap 1978). This indicates proliferation and division of the precursor cell (determined stem cell; Blackett 1971). The postnatal peak of the ^{3}H-index indicates that the proliferative activity of the determined stem cells reaches the maximum at this time. With no increased demand for thrombocytes it diminishes quickly, this being also recognizable from the parallel course of the mitotic index. Marien and McFadden (1968), however, found a continuous rise in the mitotic frequency of megakaryocyte nuclei in the rat spleen until the 40th day, a marked drop in the number of cells also being present. By contrast we could not detect any mitotic figures in megakaryocytes after 3 weeks. This discrepancy cannot be explained. But it should be emphasized that, coincident with a high ^{3}H-TdR labeling index, a high mitotic activity was found in the present investigation. Furthermore, after 40 days the hematopoietic activity of the spleen had been significantly reduced, as had also the numbers of megakaryocytes. On the other hand, the findings concerning the megakaryocyte concentration/unit area compare well with the corresponding results of Marien and McFadden (1968). In both investigations the maximal values were found at the 14th day of life.

6.3 Cell-Kinetic Studies on the Red Pulp of Newborn Rats

As far as the proliferative activity is concerned, there are two differently behaving cell groups in the red pulp at the time of birth. First there are cell populations with a fast declining radioactive labeling, and a fast turnover: the basophilic blasts, erythropoietic and myelopoietic cells, megakaryocytes, and lymphocytes. Other cells are still labeled 4 weeks after the initial administration of ^{3}H-thymidine, thus presenting a slow turnover: nonphagocytizing and phagocytizing reticulum cells and endothelial cells.

6.3.1 Basophilic Blasts

Basophilic blasts exhibit the highest proliferative activity of all cell populations in the spleen, the levels corresponding to findings for the bone marrow of rodents (Moffat et al., 1967; Miller and Osmond 1973). There is no great difference between the initial labeling indices of large and small blasts (61.2% and 74.9% respectively).

In contrast to conditions in adult animals, the large basophilic blasts in newborn rats were not 100% labeled (Yamashita and Helpap, 1974b).

By cell division and possibly immigration of labeled cells from preceding compartments, the labeling index increases to nearly 100% within 48–72 h. Then the radioactivity is rapidly diluted.

In the bone marrow of adult animals the cell cycles of basophilic and "transitional" cells amount to 15 h in vivo and 12.3—24 h in vitro, the S-phase having durations from 3.5—10.9 h depending on cell diameter (Rosse 1976). Comparable results for the extramedullary hematopoiesis in newborn rats are not available. However, t_c and t_s were determined on fetal mice just before birth (5.5 and 4.5 h respectively; Tarbutt and Cole 1970); these results should come close to the values in the newborn animal.

6.3.2 Erythropoiesis

There are no data on the proliferative activity of erythroblasts in the spleen of newborn rats. The results for the bone marrow do not allow a comparison because the first hematopoietic cells appear only 24—30 h (sternum) and 52 h (metacarpal-, metatarsal bones) after birth (Cuda, 1970). Before the 4th day the granulopoiesis is predominant with hardly any erythropoietic cells present; the peak of the erythropoietic activity is reached between the 2nd and 6th week (Reincke, 1962; Lucarelli et al. 1966). This underlines the importance of the spleen as an erythropoietic site in the postnatal period.

The erythroblasts in the spleen show a fast turnover rate (Fig. 31). One hour after birth 54.1% of the basophilic erythroblasts are labeled; the subsequent more moderate increase and plateau-like course indicate the origin of these cells from precursor cells with similar proliferative characteristics. The ^{3}H-indices of the basophilic erythroblasts and the small basophilic blasts not only show almost equal initial values, but pursue the same course. Hence it follows that the inhomogeneous group of the basophilic blasts contains proerythroblasts or their precursors (determined erythropoietic stem cells) which cannot be indentified morphologically (Starling and Rosse, 1976). On the whole, the erythroblasts in the neonatal spleen exhibit the same behavior as the erythroblasts in the bone marrow (Starling and Rosse, 1976; Odartchenko et al. 1971).

The Proliferative Pattern of Spleen Erythroblasts in 1-h-old Rats. The following analysis of the cell-kinetic parameters of the erythroblast maturation is based on the "random model", as it was applied by Tarbutt and Blackett (1968), Roylance (1968), and Constable and Blackett (1972). Provided the duration of the erythroblast cycle is known, it is possible to estimate the following parameters: rates of cell production and flow, number of cell divisions and maturation times (transit times).

Cell Cycle Times of Neonatal Spleen Erythroblasts. There is no information about the generation times of erythroblasts in neonatal rats. An evaluation from labeled mitoses curves was also impossible in the present investigation, because the examination at 6-hourly intervals exceeded the duration of the S-phase. However, t_s and t_c can be calculated from the ^{3}H-TdR and mitotic index and the known mitotic duration (Table 7). Accordingly the cell cycle of erythroblasts in the spleen of newborn rats amounts to 4.0 h.

Determination of the Number of Cell Divisions and the Transit Times in Neonatal Erythroblasts. The percentages of the different erythropoietic compartments are listed on p. 36. The proportion of proerythroblasts which cannot be identified morphologically was taken as half the percentage of the basophilic erythroblasts, because the ratio between the different erythropoietic stages is 1:2:4:8 (see above).

In exponential growth the specific birth rate (α), i.e., the cellular net gain per hour, can be calculated for erythropoietic cells as follows (Tarbutt and Blackett, 1968; Maurer and Schultze, 1968; Hanna and Tarbutt, 1971):

$$\ln 2/t_c = 0.693/4.0 \text{ h} = 0.17/\text{h} \tag{1}$$

The hourly cellular flow from the proerythroblast compartment (0.08N ; N = total amount of erythropoietic cells/animal; Constable and Blackett, 1972) amounts to:

cell number $\times$ specific birth rate, or

$$0.08\text{N} \times 0.17/\text{h} = 0.014\text{N}/\text{h} \tag{2}$$

and corresponds to the inflow into the subsequent compartment of the basophilic erythroblasts.

The cellular outflow from the population of the basophilic erythroblasts per hour can be calculated according to the equation:

outflow rate = inflow rate + ($\alpha \times$ cell number)
$$= 0.014\text{N}/\text{h} + (0.17/\text{h} \times 0.19\text{N}) \tag{3}$$
$$= 0.46\text{N}/\text{h}$$

The number of divisions of the basophilic erythroblasts is given by the ratio of inflow to outflow:

$$2^m = 0.046\text{N}/\text{h}/0.014\text{N}/\text{h}; m = 1.7 \tag{4}$$

The transit time (T) of the basophilic erythroblasts is:

$$T = m \times t_c = 6.8 \text{ h} \tag{5}$$

The outflow rate from the compartment of polychromatic erythroblasts is calculated as:

inflow rate (from basophilic erythroblasts) + ($\alpha \times$ cell number)

$$= 0.046\text{N}/\text{h} + (0.17\text{h} \times 0.33\text{N})$$
$$= 0.102\text{N}/\text{h}$$

$$2^m = 0.102\text{N}/\text{h}/0.046\text{N}/\text{h}; m = 1.2 \tag{6}$$

$$T = 1.2 \times 4.0 \text{ h} = 4.8 \text{ h}$$

The transit time of the nondividing orthochromatic erythroblasts is calculated from the ratio of the cell number to the cellular outflow from the preceding compartment (Roylance, 1968; Constable and Blackett, 1972):

$$T = 0.40N/0.102N/h = 4.0 \text{ h} \tag{7}$$

This result is somewhat lower than the transit time of 6 h that is evident in Fig. 32 (time interval between appearance of orthochromatic erythroblasts and expelled radioactively labeled erythroblast nuclei). This means that there is no ineffective erythropoiesis in the neonatal spleen. The same applies to the bone marrow erythropoiesis in young dogs (Odartchenko et al., 1971; Parry and Blackett, 1972).

The transit time (maturation time) of basophilic and polychromatic erythroblasts in the neonatal rat spleen is shorter by about the factor 3–4 than in grown-up rats, while in orthochromatic erythroblasts it remains nearly the same. The cell cycle time of the proliferating erythroblasts is markedly reduced in newborns, whereas the mitotic rate remains fairly constant as compared with adult animals. The effect of a shortened cell cycle on the divisional activity in the proliferative pool is dependent on the extent of the alterations in the maturation times (Hanna et al. 1969). If the cycle time is more reduced than the maturation time or the maturation time remains unchanged, more divisions are permitted to occur (Lord, 1967). This effect has been shown to occur after acute bleeding (Hanna, 1968). However, if maturation time and cell cycle time are shortened to the same extent, extra cell divisions cannot take place. Hence the mitotic rate in the spleen of neonatal rats is not increased. A similar situation is found with protracted anemia. Since here maturation and cell cycle time are shortened by the same proportions, the mitotic frequency does not rise. In this case the compensation occurs obviously by an increased inflow from the stem cell pool (Tarbutt, 1969; Hanna et al. 1969; Dörmer 1978). The proliferative parameters of the erythropoiesis in the neonatal rat spleen imply a mechanism similar to that in chronical anemia. There is indeed − as was already mentioned − a moderate hypochromic, macrocytic anemia in neonatal rats (Lucarelli et al., 1964, 1966).

6.3.3 Myelopoiesis

From the initial ^{3}H-index and the relative size of the various compartments (Table 10), the number of cell divisions and the transit times of granulocytic precursors can be calculated, provided that the spleen granulopoiesis does not basically differ from the bone marrow granulopoiesis and the cell cycle times of promyelocytes and myelocytes are known. T_c is 8.4 h (Table 8). Under exponential growth the granulopoietic precursors proliferate at the specific birth rate α, which is again ln $2/t_c = 0.693/8.4$ h = 0.08h (8)

The cellular outflow from the promyelocyte compartment is:

outflow rate = cell number × specific birth rate
= 0.23N × 0.08/h   (9)
= 0.018N/h

(0.23 = 23% = relative amount of promyelocytes 1 h after birth; N = total number of myelopoietic cells/animal).

The outflow rate from the myelocyte compartment is calculated from the equation:

outflow rate = inflow rate + ($\alpha \times$ cell number)
= 0.018N/h + (0.08/h $\times$ 0.49N)   (10)
= 0.057N/h

The cell division (m) in the myelocyte compartment amounts to:

2^m = outflow rate/inflow rate (Constable and Blackett, 1972),

2^m = 0.057N/h/0.018N/h = 3.3; m = 1.7   (11)

The transit time (T) in the myelocyte compartment is:

T = 1.7 $\times$ 8.4 = 14.5 h   (12)

From the initial labeling index (43%) and the hourly inflow into the myelocyte compartment (2.2%) the transit time for myelocytes is calculated to be 19.5 h if steady state growth is assumed. Within this time the labeled myelocytes have turned over once (cf. Messner, 1967). Both results stay much below the transit time of 32 h found by Constable and Blackett (1972) in the bone marrow of adult animals; in this case the mitotic activity was 2.7.

If there is no ineffective granulocytopoiesis [as was observed by Patt et al. (1964) in the bone marrow of dogs], the outflow from the myelocyte compartment equals the flow through the postproliferative stage of the metamyelocytes (Constable and Blackett 1972; Messner 1967). Thus the maturation time of the metamyelocytes can be estimated from the relative cell number and the outflow rate of the myelocytes:

T = 0.28N/0.057N/h = 5.0 h   (13)

As shown in Fig. 33, the first radioactively labeled granulocytes appear 6 h after the labeled metamyelocytes. However, this interval may well have been shorter, since only 6-h intervals were used in this investigation for the determination of the labeling index. The transit time of 5–6 h in newborn rats is considerably shorter than that obtained from adult animals (12–16 h after Constable and Blackett, 1972).

Immediately after birth, the spleen and liver are the only sites with an active hematopoiesis in rats. At birth, there are no immature granulopoietic or erythropoietic precursors present in the bone marrow, but a large number of mature normoblasts and some granulocytes are observed, which must have immigrated from extramedullary sites. This means that the proliferating myelocytes and accumulated metamyelocytes and granulocytes in the spleen are almost exclusively produced in this organ. Admittedly, some of the granulocytes may have immigrated from the liver, in which case, on the other hand, a similar proliferative pattern may be expected. In contrast to data obtained from the newborn rats, the number of divisions in the bone marrow myelocyte compartment is 2.7 in adults, and the transit time of myelocytes is 32 h while that of metamyelocytes is 12—16 h (Constable and Blackett, 1972). In the perinatal rat, the granulocytopoiesis is characterized by a reduced division rate (1.6-fold) of myelocytes and much shorter transit times (myelocytes 2.3-fold; metamyelocytes 2.4—3.2-fold). The undisturbed granulocytopoiesis in adult rats is obviously maintained by an increased mitotic rate of the visible precursor cells. In the newborn rat, however, the increased demand of granulocytes is most probably answered by a still elevated input from the determined stem cell pool (Metcalf and Stevens, 1972; Cronkite and Vincent 1969).

Finally, it should be mentioned that the extramedullary myelopoiesis in the spleen of adult rats has almost completely ceased, (as has splenic hematopoiesis in general), and a direct comparison of the myelopoiesis in this organ with that of newborn rats is not possible. The decline of the splenic myelopoietic activity takes place between the 5th and 10th day of life (Lord 1965a) and is closely related to a gradual transition from an actively proliferating to a more resting stem cell pool, and probably also to developmental changes in the hematopoietic-inductive microenvironment.

6.3.4 Lymphocytes

From the absence of labeled lymphocytes 1 h after ^{3}H-TdR injection, it can be concluded that there is obviously no proliferation of lymphocytes in the red pulp at this time (Köbberling, 1965; Hinrichsen, 1967). The first labeled lymphocytes which appear after 12 h must have incorporated ^{3}H-TdR somewhere else (bone marrow, thymus); they reach the red pulp by the blood stream after a maturation and migration time of at least 12 h. Because of different rates of inflow, two populations of lymphocytes must exist: one remains in the red pulp for a maximum of 22 h, the other for about 5 days. Taking a certain loss by cell death and transformation into plasmocytoid cells into account, the actual stay in the red pulp should be somewhat shorter.

Immediately after birth, spleen and lymph nodes in rodents (rats, mice) are still immature; they are populated to an increasing extent by the primary lymphatic organs – bone marrow and thymus – which assume a special role (Joel et al. 1972). Initial T cell functions can be demonstrated in the mouse spleen on the 3rd—4th postnatal day (Chiscon and Golub, 1972; Hardy et al. 1976; Friedberg and Weißmann 1974), and the rat spleen only from the 14th day on (Veerman, 1975). However, functioning B cells which are able to produce IgM are already present 24—48 h postnatally (Friedman and Globerson 1975; Hardy et al. 1976; Mosier and Johnson 1975).

In newborn mice (18—30 h old) ^{3}H-TdR labeled thymus lymphocytes appeared 24 h after the thymic labeling in the spleen, where they gathered in the periarteriolar zone and only occasionally in the red pulp. At this time 26% of the spleen lymphocytes originated from the thymus (Joel et al. 1972).

As far as the nature of the lymphocytes in the red pulp of newborn rats is concerned, only presumptions can be made (Grouls and Helpap 1979). They seem to be mainly B-lymphocytes, because the B cell activity can be demonstrated at an earlier stage than the T cell activity. Furthermore, the lymphocytes in the spleen red pulp of adult animals are predominantly thymus independent (deSousa and Parrott 1967; Parrott and deSousa 1971). The long residency of the labeled lymphocytes in the neonatal spleen (up to 5 days) also stands more in favor of their B cell nature. Though B-lymphocytes do recirculate, their motility is much lower than that of T-lymphocytes, which need only 5—6 h to pass the spleen in grown-up animals (Sprent 1973; Ford 1975; 1969).

Not much is known about the recirculation of B-lymphocytes; they are difficult to mobilize and about 10 days are necessary to achieve a depletion of the B cell zones in the spleen over a drainage of the thoracic duct (Sprent 1973).

6.3.5 Reticulum Cells

The data on the frequency of the reticulum cells in the spleen of newborn rats differ considerably due to different techniques (paraffin sections, smears) and different definitions. Including histiocytes and mesenchymal cells ("large, undifferentiated cells") within the group of reticulum cells, Köbberling (1966) found 45% of these cells. On the other hand, Riman et al. (1958) observed only 0.47% of reticulum cells in smears of homogenated spleen tissue. We found 22.2% of nonphagocytizing and 4.2% of phagocytizing reticulum cells.

The initially high percentage of radioactively labeled reticulum cells and macrophages (11.6% and 5.5% respectively) indicates a high local proliferation of these cells. The further increase of the ^{3}H-indices until the 5 th day can be interpreted as follows:
1. It is an expression of an increased mitotic activity and reutilization, especially of the macrophages.
2. Cells which were labeled somewhere else before, immigrate and settle down in the spleen as phagocytizing or nonphagocytizing reticulum cells.

However, turnover and mitotic activity should be low, since the labeling is still demonstrable between the 7th and 28th day (Caffrey et al. 1966). The subsequent fall of the ^{3}H-indices could be explained by the immigration of nonlabeled reticulohistiocytic cells and the effect of dilution (Haas et al. (1972).

According to recent investigations, both reticulum cells and macrophages are derived from immigrated monocytes (van Furth and Thompson 1971; Roser, 1970). These cells originate from promonocytes in the bone marrow; in contrast to the latter they have lost the ability to synthesize DNA (van Furth and Diesselhoff den Dulk, 1970). In mice labeled monocytes were demonstrated in the peripheral blood 2 h after the injection of ^{3}H-TdR. From this observation it was concluded that monocytes leave the bone marrow immediately after their formation (van Furth and Cohn 1968). As soon as the monocytes have reached the tissue they show marked phagocytic activity and are called tissue macrophages (van Furth and Thompson 1971). Under normal conditions tissue macrophages exhibit no or only a very low mitotic activity, which, however, may be raised by stimulating influences like inflammations (Roser 1970).

6.3.6 Monocytes

There are only few monocytes in the neonatal splenic red pulp: 0.4% were found in this study, 0.2% by Riman et al. (1958). Radioactively labeled monocytoid cells could not be found in semithin sections. This might be due to the transformation of monocytes into reticulum cells and macrophages as soon as they have reached the spleen. Some of them probably remain in the periarteriolar zone of the future white pulp as precursors for the specific interdigitating cells, which are characteristic of this area (Veerman, 1974, 1975; Wiersbowsky et al., 1982).

55

6.3.7 Endothelial Cells

The number of endothelial cells in newborn rats is only known for the bone marrow (3.1% ± 1.3%; Haas et al. 1967). In the spleen of adult mice 3.3 ± 1.4% were found (Yamashita and Helpap 1974b). Yamashita and Helpap (1974d) observed 16.9% ± 3.3% sinus endothelial cells in the spleen of just grown-up rats; 1 h after ^{3}H-thymidine injection these cells were not labeled.

In newborn rats we determined 5.3% of endothelial cells; the labeling index of 11.2% 1 h after the injection of ^{3}H-TdR indicates a significant local proliferation of these cells. The subsequent rise of the labeling indices until the 5th day can be interpreted as follows:

1. In the still immature spleen a rapid turnover of these cells is present which is regulated and maintained by the body and organ growth and the increase of the circulating blood volume.
2. Reutilization occurs, i.e., already labeled DNA compounds from perished cells are incorporated again.
3. Cells which were labeled somewhere else (bone marrow) immigrate into the neonatal and transform into (sinus-) endothelial cells.

Endothelial cells are a cell population with a low turnover rate, because a lot of them are still labeled after 4 weeks. Bone marrow endothelial cells which were labeled to 100% at the time of birth, showed an ^{3}H-index of 40%–50% after 10 weeks and remained at this level for the following 5 months (Haas et al. 1967, 1969; Fliedner et al. 1968a, b). Determinations of the silver grain density showed that the labeling intensity decreased by 50% within the first 24–30 h. But this initial decrease was not sufficient to make the percentage of labeled endothelial cells drop below 100% within the first few days (Fliedner et al. 1968a, b). If the results for the bone marrow are transferred to the spleen, it can be presumed that endothelial and stromal cells have divided at least once within 24–30 h; the ^{3}H-index then stays on the same level. The increase of the labeling index until the 5th day can only be explained by an immigration of transformed cells by the blood stream and which were labeled before, e.g., in the bone marrow. It should be noted here that the group of "endothelial and stromal cells" comprises partly heterogeneous cell populations, i.e., endothelial cells of small venules and arterioles, sinus endothelial cells, and probably partly fibroblastic and fibrocytic cell elements. Only the sinus endothelial cells and fibroblasts are of a possible hematogenous origin (Büchner 1971; Stutte 1974; Helpap and Cremer 1972; Pictet et al. 1969).

Monocytes are considered the precursors of sinus endothelial cells in the spleen and bone marrow and Kupffer cells in the liver (van Furth and Thompson 1971; Warr and Sljivic 1974; Nicolescu and Roullier 1967; Stutte et al. 1974). They should – under the influence of the microenvironment – also be able to transform postnatally into sinus endothelial cells, fibroblasts (?), reticulum cells, and tissue macrophages.

6.3.8 Megakaryocytes

The percentage of megakaryocytes in the spleen of newborn rats ranges between 0.03% (Riman et al. 1958) and 0.1% (Lucarelli et al. 1966). The amount in semithin sections is 0.5% ± 0.17%. From the three different types of megakaryocytes only type

I is still capable of synthesizing DNA. The fast increase in the labeling index to 62.2% after 21 h indicates a fast transit time of the precursor cell. (Blackett, 1971; Yamashita and Helpap 1974c; Grouls and Helpap 1978). On the whole the ^{3}H-indices of mega-karyocytes in the neonatal rat spleen (Fig. 37) pursue the same course as in grown-up animals (Ebbe and Stohlman 1965). From the increase in labeling index within 21 h (2.9%/h) the transit time — the time when all the labeled megakaryocytes are replac-ed by nonlabeled ones — can be determined. This transit time (from the precursor cell to the megakaryocyte type III) amounts to 34.5 h; it is much shorter than the transit time of 57 h in the mouse spleen (Yamashita and Helpap 1974c). The transit times of megakaryocytes in the bone marrow of grown-up rats and mice are 43—75 h (Ebbe and Stohlman 1965) and 50—57 h (Odell et al. 1969) respectively. Thus at the time of birth megakaryocytopoiesis in the spleen is — similar to the erythro- and myelopoie-sis — supplied by an increased introduction of precursor cells into the megakaryocyte compartment (Grouls and Helpap 1980).

7 Closing Remarks

Cell-analytical and cell-kinetical studies on hematopoietic and stromal cells in the red pulp of rats of different age groups were carried out by means of semithin sections and semithin section autoradiographs. Due to a high light-optical resolution, this me-thod allows a reliable cell classification, in contrast to stripping film autoradiographs from paraffin sections.

The normal values determined shall serve for the evaluation of the spleen under the influence of inflammatory, immunological, and tumorous processes.

In the rat three organs are involved in hematopoiesis at the time of birth: liver, spleen, and bone marrow. Whereas hematopoiesis in the liver ceases within 7 days after birth and is only gradually intensified in the bone marrow, it increases significant-ly in the spleen within the first weeks of life. At the time of birth the still immature bone marrow which contains mainly granulopoietic cells is not yet able to cover the increasing demand of blood and to replace the normal loss of cells. The spleen is also involved in the production of granulocytic cells, the demand of which is high because of the initial contact with infectious agents and a still immature lymphatic cell system.

Until the 3rd day the increased hematopoiesis in the spleen is accomplished by a high turnover in the stem cell pool, and until the 5th day by an additional shortening of the generation times of the immature hematopoietic cells. Then the generation times become prolonged again; the stem cells turn into a more resting, barely differen-tiating pool. When these alterations in the spleen occur, the hematopoietic activity in the bone marrow reaches its maximum. At the same time a hypochromic, macrocytic, partly hemolytic anemia persists and ceases completely only after 2 months. At birth the rat has a reticulocytosis of about 90%, by the 20th day of 20%, and only by 60 days are normal adult values of 2%—3% reticulocytes and hematocrits of 40%—48% reached (Stohlman et al. 1964). The early reduction of the compensatory spleen hema-topoiesis does not agree with these hematologic findings in the peripheral blood. The reason for the early decrease in the extramedullary hematopoiesis is unknown; it might be due to different properties of fetal and adult hematopoietic stem cells. The transition from one stage to the other takes place only slowly and a period of relative hypoplasia may occur when none of the two systems operates with full capacity (Stohlman et al. 1964).

A postnatal absolute or relative iron deficiency might also affect the activity of the stem cells, since suppressed differentiation of determined stem cells was considered the basic mechanism in iron deficiency anemias of different etiology (Dörmer and Lau 1977, 1978). The iron requirements are particularly large and rapid during the replacement of embryonic red cells and formation of the first adult red cells (Theil, 1980). However, in adults iron deficiency evokes a characteristic microcytic hypochromic anemia, contrasting with the macrocytic pattern in newborn rats. This unusual character of the red cells in newborn rats was taken as evidence of different stem cell properties in these animals (Stohlman et al. 1964; Lucarelli et al., 1967).

Alterations similar to these described in the present investigation should occur in the human spleen and bone marrow, since the hematopoiesis in human embryos of the 5th—7th gestational months corresponds to that in newborn rats. Thus congenital hypo- and aplasias of erythropoiesis could be derived from an impaired transition from the fetal to the adult erythropoiesis (Stohlman et al. 1964).

The erythropoiesis in the spleen which occurs under pathological conditions (e.g., hemorrhages) starts from the few large basophilic blasts which are still present in the adult spleen (^{3}H-index 80%—100%) by increasing the cellular inflow from the preceding stem cell pool (Renricca et al. 1970). On the other hand, the generation times of erythroblasts shorten rapidly.

For an effective proliferation and turnover of hematopoietic cells an intact environment is necessary, the so-called hematopoietic-inductive microenvironment (McCuskey et al. 1972; Trentin 1971). Whereas the stem cell pool cannot be exhausted, age-dependent alterations (arteriosclerosis, interstitial and periadventitial fibrosclerosis, hyalinization, and arterial microthrombosis) affect the microenvironment and may cause a decrease in the hematopoiesis (Fliedner and Heit 1975). A significant hyalinosis in the spleen is already be found at an early age in humans and causes impairment of microcirculation and oxygen supply with subsequent alterations of the mesenchymal ground substances (mucopolysaccharides). Alternatively and/or additionally the renewal of the hematopoietic microenvironment by the immigration of monocytes and their transformation into specialized macrophages might also be disturbed (Ploemacher and van Soest 1977a, b).

The extramedullary hematopoiesis which occurs under pathologic conditions (myeloproliferative syndromes like osteomyelofibrosis and chronic myeloic leukemia, severe hemolytic anemia; tumor anemia, especially bone marrow carcinosis) in humans is bound to take place in an altered unfavorable microenvironment (Löffler, 1971; Söderström et al. 1975). Therefore, this compensatory hematopoiesis can no longer be effective (Hoelzer and Harriss, 1973; Neumann et al. 1976; Söderström et al. 1975; Sjögren 1978, 1979; Sjögren and Brandt 1976).

The present results should be regarded as a starting point for further investigations on extramedullary stem cells and heterotopic hematopoiesis after experimental stimulation or various hematologic diseases (hypoplasia, aplasia, leukemia). The spleen is also the most suitable organ for the examination of the relationship between immunologic pathology and hematopoiesis. The method of semithin sections and semithin section autoradiography has proved very useful, and might also be facilitated in the evaluation of age-dependent cellular and kinetic hematopoietic changes in human bone marrow biopsies.

8 Summary

The extramedullary hematopoiesis and the development of stromal cells in the red pulp of rats from birth until 60 days of age were investigated by means of semithin sections and semithin section autoradiographs. Thirty-eight Wistar rats of both sexes were used. They were killed at various time intervals after birth (1 h, 18 h, 3, 5, 10, 14, 20, 30, 45 and 60 days). One hour prior to death each animal received an intraperitoneal injection of 3 μCi ^{3}H-thymidine (^{3}H-TdR)/g body weight.

Furthermore, 33 newborn Wistar rats were examined; these rats were administered an intraperitoneal injection of 3 μCi ^{3}H-TdR 1 h after birth and were killed 1, 6, 12, 18, 21, 44 h and 3, 5, 7, 10, 12, 14 and 28 days later. The specific activity of ^{3}H-TdR amounted to 20 Ci/mMol. From each spleen eight to ten tissue pieces (1 x 1 mm) were embedded into Epon and then cut into 0.8–1 μm thick sections using the LKB III ultramicrotome. Autoradiographs were obtained by the dipping technique. The sections were stained with basic fuchsin and methylene blue azure II. In the histologic sections and autoradiographs the cells were analyzed (percentage distribution), and the cell number per unit area (0.01 mm^2) and ^{3}H-TdR and mitotic indices were determined. The following hematopoietic and stromal cells could be distinguished: small and large basophilic blasts; basophilic, polychromatic, and orthochromatic erythroblasts; promyelocytes, myelocytes, metamyelocytes, and granulocytes; lymphocytes; plasma cells; monocytoid cells; megakaryocytes; phagocytizing and nonphagocytizing reticulum cells as well as endothelial cells. The following results were obtained:

The spleen hematopoiesis in newborn rats is mainly supplied by an active stem cell pool, thus resembling the hematopoiesis at the end of the fetal life. An ineffective spleen erythropoiesis does not occur at this stage. The transit times of basophilic, polychromatic, and orthochromatic erythroblasts amount to 6.8, 4.8, and 4.0 h. No indications for an ineffective granulopoiesis were found. Some of the myelocytes divide a second time. Their transit time is 14.5 h, while the transit time of the metamyelocytes is 5.0 h.

There are obviously two populations of lymphocytes in the red pulp which differ in the times of residency in the spleen. Phagocytizing and nonphagocytizing reticulum cells and endothelial cells derive obviously from immigrating cells (monocytes). The transit time of megakaryocytes from the precursor cell to the megakaryocyte type III amounts to 34.5 h, which is much shorter than in grown-up animals.

Since after 2 months a small amount of DNA-synthesizing basophilic blasts is still labeled to 90%–100%, the hematopoiesis in the spleen can presumably be increased to a considerable extent under pathologic conditions, even in older animals. Then the rapidly activated extramedullary hematopoiesis originates from the above-mentioned "immature" basophilic blasts with stem cell properties.

A significant extramedullary erythropoiesis was observed until an age of more than 30 days, whereas the results on smears imply a much lower activity.

As far as the relationship between the different erythroblasts is concerned, the polychromatic erythroblasts have a decreased mitotic activity and turn prematurely into orthochromatic erythroblasts. Macrocytes are formed which are typical of the peripheral blood in the rat in the postnatal stage.

From the pattern of the mitotic and labeling indices it becomes evident that the generation times and S-phases of the proliferating hematopoietic cells become longer with increasing age. Within 5—10 days after birth a gradual transition from an actively proliferating to a more resting stem cell pool occurs; cell production in the hematopoietic proliferation pool is reduced. Different regulating mechanisms and stem cell properties in the fetal and adult hematopoiesis might be responsible for the decrease in the proliferation in the presence of the postnatal anemia. During the transitional stage obviously a relative hypoplasia occurs with neither of the two mechanisms fully functioning.

As the spleen is an integral part of the hematopoietic cell system in rats and a possible compensatory organ for the bone marrow, the alterations in the spleen represent the physiologic principles of bone marrow hematopoiesis. Besides, the human fetal hematopoiesis should develop in a similar way, since hematopoiesis in newborn rats resembles hematopoiesis in the human fetus in the 5th—7th month of gestation.

Acknowledgments. The valuable assistance of Miss Inge Heim and Dr. Dorothea Mikolai is gratefully acknowledged.

60

References

Amlacher E (1974) Autoradiographie in Histologie und Zytologie. Thieme, Leipzig

Andrew W (1946) Age changes in the vascular architecture and cell content in the spleens of 100 Wistar Institute rats, including comparisons with human material. Amer J Anat 79:1–74

Baserga R, Malamud D (1969) Modern methods in experimental pathology. Autoradiography. Techniques and application. Hoeber Medical Division, Harper and Row, New York Evanston London

Baserga R, Wiebel F (1969) The cell cycle of mammalian cells. Int Rev Exp Path 7:1–27

Bessis M (1973) Living blood cells and their ultrastructure. Springer, Berlin Heidelberg New York

Bessis M (1977) Blood smears, reinterpreted. Springer, Berlin Heidelberg New York

Blackett NM (1971) The proliferation and maturation of hemopoietic cells. In: Baserga R The cell cycle and cancer. Dekker, New York pp 26–53

Block M (1976) Bone marrow examination. Aspiration or core biopsy, smear or section, haematoxylin-eosin or Romanowsky stain – which combination? Arch Pathol Lab Med 100:454–456

Blümcke S, Backmann R (1966): Epon 812 als Einbettungsmittel für die lichtmikroskopische Autoradiographie. Naturwissenschaften 53:362

Boggs DR, Geist A, Chervenick PA (1969) Contribution of the mouse spleen to post-hemorrhagic erythropoiesis. Life Sci 8:587–599

Boll I (1978) Das granulocytäre Zellsystem. In: Queißer W (ed) Das Knochenmark, Morphologie, Funktion, Diagnostik, Thieme, Stuttgart pp 167–192

Bozzini CE, Barrio Rendo ME, Devoto FC, Epper CE (1970) Studies on medullary and extramedullary erythropoiesis in the adult mouse. Amer J Physiol 219:724–728

Bryant BJ (1963) Reutilization of lymphocyte DNA by cells of intestinal crypts and regenerating liver. J Cell Biol 18:515–523

Büchner Th (1971) Entzündungszellen im Blut und im Gewebe. (Zellkinetische Studie über experimentelle granulierende Entzündung durch Fremdkörper und bei der Wundheilung). Veröffentlichung aus der morphologischen Pathologie, Fischer, Stuttgart

Calvo W, Fliedner TM, Herbst EW, Fache I (1975) Regeneration of blood-forming organs after autologous leukocyte transfusion in lethally irradiated dogs. I. Distribution and cellularity of the bone marrow in normal dogs. Blood 46:453–457

Caffrey RW, Everett NB, Rieke WO (1966) Radioautographic studies of reticular and blast cells in the hemopoietic tissues of the rat. Anat Rec 155:41–58

Chiscon MO, Golub ES (1972) Functional development of the interacting cells in the immune response. I. Development of T cell and B cell function. J Immunol 108:1379–1385

Constable TB, Blackett NM (1972) The cell population kinetics of neutrophilic cells. Cell Tissue Kinet 5:289–302

Corsi A, Giusti GV (1967) Cellular distribution in the bone marrow after thymectomy. Nature 216:493–494

Cottier H, Odartchenko N, Feinendegen LE, Keiser G, Bond VP (1963) Autoradiographische Untersuchungen über die Entkernung der Erythroblasten nach In-Vivo-Markierung mit Thymidin-[3] H. Schweiz Med Wschr 33:1061–1065

Cronkite EP, Fliedner TM (1964) Granulocytopoiesis (Concluded). New Engl J Med 270:1403–1408

Cronkite EP, Vincent PC (1969) Granulocytopoiesis. Ser Haemat II:3–43

Cuda G (1970) Zytokinetische Untersuchungen über die Ontogenese des Rattenknochenmarks. Beitrag zur Frage der Haemopoeseentstehung aus ortsständigen oder zirkulierenden Stammzellen. Inaug Diss, Ulm

Culbertson JT (1939) Phagocytosis of trypan blue in rats of different age groups. Arch Pathol 27:212–217

Curry JL, Trentin JJ (1967) Hemopoietic spleen colony studies. I. Growth and differentiation. Dev Biol 15:395–413

Dicke KA, Van Noord MJ, Maat B, Schaefer UW, Van Bekkum DW (1973) Identification of cells in primate bone marrow resembling the hemopoietic stem cell in the mouse. Blood 42:195–208

Dörmer P (1973) Kinetics of erythropoietic cell proliferation in normal and anemic man. A new approach using quantitative ^{14}C-autoradiography. Fischer, Stuttgart

Dörmer P, Lau B (1977) Erythropoese bei Eisenmangel. Blut 34:453–464

Dörmer P, Lau B (1978) Kinetics of erythroblast proliferation in states of hypoferremia. Isr J Med Sci 14:1144–1151

Dörmer P (1978) Das erythrozytäre Zellsystem, In: Queißer W. (ed.): Das Knochenmark, Morphologie–Funktion–Diagnostik, Thieme, Stuttgart pp 150–167

Ebbe S, Stohlman F Jr (1965) Megakaryocytopoiesis in the rat. Blood 26:20–35

Ebbe S, Stohlman F Jr, Donovan J, Overcash J (1968) Megakaryocyte maturation rate in thrombocytopenic rats. Blood 32:787–795

Feinendegen LE, Bond VP, Hughes WL (1966) Physiological thymidine reutilization in rat bone marrow. Proc Soc Exp Biol Med 122:448–451

Feinendegen LE, Heiniger HJ, Friedrich GR, Cronkite EP (1973) Differences in reutilization of thymidine in hemopoietic and lymphopoietic tissues of the normal mouse. Cell Tissue Kinet 6:573–585

Fischer R, Hennekeuser HH, Schaefer HE (1970) Extramedulläre Blutbildung in der Milz, insbesondere bei Knochenmarkmetastasierung. In: Lennert K, Harms D (eds) Die Milz. Struktur, Funktion, Pathologie, Klinik, Therapie. Springer, Heidelberg Berlin New York pp 81–93

Fliedner TM (1974) Kinetik und Regulationsmechanismen des Granulozytenumsatzes. Schweiz Med Wschr 104:98–107

Fliedner TM (1975) Hämopoetische Stammzellen: Eine Teilpopulation der "Lymphozyten". In: Theml H, Begemann H (eds) Lymphozyt und klinische Immunologie: Physiologie – Pathologie – Therapie. Springer, Berlin Heidelberg New York

Fliedner TM (1976) Das granulozytäre Zellerneuerungssystem: Ein Regelkreis. In: Stacher A, Höcker P (Hrsg) Erkrankungen der Myelopoese. Leukämien, myeloproliferatives Syndrom, Polyzythämie. Urban & Schwarzenberg, München Berlin Wien

Fliedner TM, Cronkite EP, Bond VP (1959) Die Proliferationsdynamik der Blutzellbildung, autoradiographisch untersucht mit tritium-markiertem Thymidin. Schweiz Med Wschr 89: 1061–1082

Fliedner TM, Cronkite EP, Bond VP (1961) Das Studium der Proliferationsdynamik der Myelopoese unter Verwendung der Einzelzellautoradiographie. Folia Haematol N F 6:210–228

Fliedner TM, Haas RJ, Stehle H, Adams A (1968a) Complete labeling of all cell nuclei in newborn rats with ^{3}H-thymidine. A tool for the evaluation of rapidly and slowly proliferating cell systems. Lab Invest 18:249–259

Fliedner TM, Haas RJ, Stehle H (1968b) Die ^{3}H-Thymidinmarkierung aller Zellkerne neugeborener Ratten. (Eine Methode zur zytokinetischen Differenzierung von schnell und langsam proliferierenden Zellsystemen). Acta Histochem Suppl 8:231–256

Fliedner TM, Messner H, Kubanek B (1969) Neuere Erkenntnisse zur Physiologie und Pathophysiologie der Erythropoese. Haematol Bluttransfus 8:1–15

Fliedner TM, Calvo W (1969) Orthologie und Pathologie der Knochenmarksregeneration. In: Hdb Allg Path Bd VI/2, Regeneration, Hyperplasie, Cancerisierung. Springer, Berlin Heidelberg New York

Fliedner TM, Heit H (1975) Altersvorgänge in Zellerneuerungssystemen. Verh Dtsch Ges Pathol 59:71–77

Ford WL (1969) The immunological and migratory properties of the lymphocytes recirculating through the rat spleen. Brit J exp Path 50:257–269

Ford WL (1975) Lymphocyte migration and immune responses. Prog Allergy 19:1–59

Fresen O (1960) Orthologie und Pathologie der heterotopen Hämopoese. Ergeb Allg Pathol Pathol Anat 40:139–198

Friedberg SH, Weissmann IL (1974) Lymphoid tissue architecture II. Ontogeny of peripheral T and B cells in mice: evidence against Peyer's patches as the site of generation of B-cells. J Immunol 113:1477–1492

Friedman D, Globerson A (1975) Development of Ig M and Ig G antibody response in newborn and young mice. Israel J Med Sci 11:1376

Fritsch H, Queißer W (1978) Auswertung und Beurteilung des Knochenmarkes. In: Queißer W (Hrsg) Das Knochenmark. Morphologie–Funktion–Diagnostik. Thieme, Stuttgart pp 19–29

Furth VR van, Cohn ZA (1968) The origin and kinetics of mononuclear phagocytes. J Exp Med 128:415–435

Furth VR van, Diesselhoff-Den Dulk MMC (1970) The kinetics of promonocytes and monocytes in the bone marrow. J Exp Med 132:813–828

Furth VR van, Thompson J (1971) Review of the origin and kinetic of the promonocytes, monocytes and macrophages and brief discussion of the mononuclear phagocyte system. Ann Inst Pasteur 120:337–335

Gerecke D, Gross R (1976) Autoradiographic evidence for reutilisation of DNA catabolites by granulocytopoiesis in the rat. Scand J Haematol 17:132–142

Grouls V, Helpap B (1978) Influence of partial hepatectomy on the megakaryocytopoiesis of the spleen. Res Exp Med (Berl) 173:245–249

Grouls V, Helpap B (1979) Das postnatale Verhalten der Lymphocyten in der roten Milzpulpa von Ratten. Verh Dtsch Ges Pathol 63:544

Grouls V, Helpap B (1980) Megakaryocytopoiesis in the spleen of growing rats. Am J Anat 157: 429–432

Haas RJ, Stehle H, Fliedner TM (1967) Autoradiographic studies on rapidly and slowly proliferating cell systems in neonatal bone marrow. Helv Med Acta 34:54–66

Haas RJ, Bohne F, Fliedner TM (1969a) On the development of slowly-turning-over cell types in neonatal rat bone marrow. (Studies utilizing the complete tritiated-thymidine labeling method complemented by C-14 thymidine administration). Blood 34:791–805

Haas RJ, Fliedner TM, Sparrer E (1969b) Autoradiographische Untersuchungen zur zytokinetischen Analyse der neonatalen Erythropoese bei Ratten. Hämatol Bluttransfus 8:21–25

Haas RJ, Sparrer E, Fliedner TM (1970) Zur Proliferationskinetik der Lebererythropoese heranwachsender Ratten. Acta Haematol 43:232–241

Haas RJ, Meyer-Hamme KD, Trepel F, Frölich D (1972) Cytokinetic studies on slowly-renewing bone marrow and spleen lymphocytes in rats during a primary and secondary immune response. Acta Haematol 48:39–48

Hanna IRA (1967) Response of early erythroid precursors to bleeding. Nature 214:355–357

Hanna IRA (1968) An early response of the morphologically recognizable erythroid precursors to bleeding. Cell Tissue Kinet 1:91–100

Hanna IRA, Tarbutt RG (1971) The relationship between cell maturation and proliferation in the erythroid system of the rat. Cell Tissue Kinet 4:47–59

Hanna IRA, Tarbutt RG, Lamerton LF (1969) Shortening of the cell-cycle time of erythroid precursors in response to anaemia. Brit J Haematol 16:381–387

Hardy J (1967) Haematology of rats and mice. In: Cotchoi E, Roe FJC (eds) Pathology of laboratory rats and mice, Blackwell, Oxford Edinburgh, pp 501–536

Hardy B, Mozes E, Danon D (1976) Comparison of the immune response potential of newborn mice to T-dependent and T-independent synthetic polypeptides. Immunol 30:261–265

Harris C, Burke WT (1957) The changing cellular distribution in bone marrow of the normal albino rat between one and fifty weeks of age. Am J Path 33:931–951

Haurani FJ, Tocantins LM (1961) Ineffective erythropoiesis. Am J Med 31:519–531

Heiniger HJ, Feinendegen LE, Bürki K (1971) Reutilization of thymidine in various groups of rat bone marrow cells. Blood 37:340–348

Helpap B, Cremer H, Grouls V (1971) Reutilisation von markierten DNS-Bausteinen bei der Wundheilung. Naturwissenschaften 58:574–575

Helpap B, Cremer H (1972) Autoradiographische Untersuchungen am Granulationsgewebe mit radioaktiv markiertem und unmarkiertem Thymidin. Virchows Archiv Zellpathol 10:145–151

Helpap B, Dachselt U (1978) The pattern of lymphocytes in the thymus and spleen after labeling with ^{3}H-thymidine and ^{3}H-deoxycytidine. Virchows Archiv [Cell Pathol] 28:287–299

Hennekeuser HH, Fischer R (1967) Extramedulläre Blutbildung und leukämoide Reaktion bei bösartigen Tumoren. Dtsch Med Wschr 92:479–482

Hinrichsen K (1967) Thymidine ^{3}H in developing germinal centers. In: Cottier H, Odartchenko N, Schindler R, Congdon CC (eds) Germinal centers in immune response. Springer, Berlin Heidelberg New York

Hoelzer D, Fliedner TM, Harriss EB, Queißer W (1972) Umsatzkinetik der Erythropoese bei „Erythroleukämie". In: Gross R, van den Loo J (eds) Leukämie. Springer, Berlin Heidelberg New York pp 381–386

Hoelzer D, Harriss EB (1973) The failure of normal haemopoiesis in rats during the development of acute leukaemia. Acta Haematol 49:36–47

Hoelzer D, Kurrle E, Harriss EB, Fliedner TM, Haas RJ (1975) Evidence for stem cell function of resting bone marrow lymphocytes identified by the complete ^{3}H-thymidine labeling method. Biomed 22:285–290

Joel DD, Hess MW, Cottier H (1972) Magnitude and pattern of thymic lymphocyte migration in neonatal mice. J Exp Med 135:907–923

Killmann SA, Cronkite EP, Fliedner TM, Bond VP (1964) Mitotic indices of human bone marrow cells. III. Duration of some phases of erythrocytic and granulocytic proliferation computed from mitotic indices. Blood 24:267–280

Killmann SA (1970) Cell classification and kinetic aspects of normoblastic and megaloblastic erythropoiesis. Cell Tissue Kinet 3:217–228

Kindred JE (1940) A quantitative study of the hemopoietic organs of young albino rats. Am J Anat 67:99–149

Kindred JE (1942) A quantitative study of the hemopoietic organs of young adult albino rats. Am J Anat 71:207–243

Köbberling G (1965) Autoradiographische Untersuchungen über Zellursprung und Zellwanderung in lymphatischen Organen fetaler und neugeborener Mäuse. Z Zellforsch 68:631–659

Langdon HL, Berman J (1975) An autoradiographic and morphological study of mouse bone marrow littoral cells during and after treatment with urethan. Cell Tissue Kinet 8:285–296

Lajtha LG (1975) Annotation: haemopoietic stem cells. Brit J Haemat 29:529–535

Leder LD (1967) Der Blutmonozyt. Experimentelle Medizin, Pathologie und Klinik. Springer, Berlin Heidelberg New York

Leibetseder F (1948) Erythropoese und Zellkerngröße. Wien Z Inn Med 29:397–408

Lennert K (1952) Zur Praxis der pathologisch-anatomischen Knochenmarksuntersuchung. Frankf Z Path 63:267–299

Lennert K (1961) Lymphknoten, Cytologie und Lymphadenitis. In: Uehlinger E (ed) Handbuch der speziellen pathologischen Anatomie und Histologie Bd I/3A Springer, Berlin Heidelberg New York

Lennert K, Müller-Hermelink HK (1975) Lymphocyten und ihre Funktionsformen-Morphologie, Organisation und immunologische Bedeutung. Verh Anat Ges 69:19–62

Lennert K (1978) Malignant lymphomas other than Hodgkin's disease. In: Uehlinger E (ed) Handbuch der speziellen pathologischen Anatomie und Histologie, Bd. I/3 B Springer, Berlin Heidelberg New York

Liebermann-Meffert D (1971) Die Erythropoese in der menschlichen Milz während des dritten bis vierten Fetalmonates. Mat Med Nordm 23:68–81

Lipkin M (1971) The proliferative cycle of mammalian cells. In: Baserga R (ed) The cell cycle and cancer, pp 6–26, Dekker New York

Löffler H (1971) Erythrozytenbildung und Erythrozytenreifung in der menschlichen Milz. Mat Med Nordmark 23:82–93

Lord BI (1965a) Haemopoietic changes in the rat during growth and during continuous gamma irradiation of the adult animal. Brit J Haematol 11:525–536

Lord BI (1965b) Cellular proliferation in normal and continuously irradiated rat bone marrow studied by repeated labeling with tritiated thymidine. Brit J Haematol 11:130–143

Lord BI (1967) Erythropoietic cell proliferation during recovery from acute haemorrhage. Brit J Haematol 13:160–167

Lord BI (1970) Measurement of the duration of DNA-synthesis in erythroid cells of the bone marrow using a double labelling autoradiographic technique designed specifically for use with haemopoietic tissues. Cell Tissue Kinet 3:13–19

Lucarelli G, Howard D, Stohlmann F Jr (1964) Regulation of erythropoiesis. XV. Neonatal erythropoiesis and the effect of nephrectomy. J Clin Invest 43:2195–2203

Lucarelli G, Porcellini A, Carnevali C, Ferrari L, Rizzoli V, Howard D, Stohlmann F, Butturini U (1966) L'emopoiesi nel periodo fetale e neonatale del ratto. L'Ateneo Parmese Acta Biomed 37:293–339

Lucarelli G, Porcellini A, Carnevali C, Carmena A, Stohlmann F Jr (1968a) Fetal and neonatal erythropoiesis. Ann NY Acad Sci 149:544−559

Lucarelli G, Porcellini A, Carnevali C, Rizzoli V (1968b) Zur Proliferationskinetik der fötalen Erythropoese der Leber. Jahresbericht 1965−1966, Europ. Atom Gem.-Gesellschaft f. Strahlenforschung EUR 938d, 47 (cited after Haas et al. 1970)

Luft JH (1961) Improvements in epoxyresin embedding methods. J Biophys Biochem Cytol 9:409−414

Maloney MA, Patt HM, Lund JE (1971) Granulocyte dynamics and the question of ineffective granulopoiesis. Cell Tissue Kinet 4:201−209

Maloney MA, Weber CL, Patt HM (1963) Myelocyte − metamyelocyte transition in the bone marrow of the dog. Nature 197:150−152

Marien GJ, McFadden KD (1968) A study of megakaryocytes in albino rats. Can J Zool 46:1053−1058

Maurer W, Schultze B (1968) Überblick über autoradiographische Methoden und Ergebnisse zur Bestimmung von Generationszeiten und Teilphasen von tierischen Zellen mit markiertem Thymidin. Acta Histochem Suppl VIII:73−87

Maximow A (1910) Untersuchungen über Blut und Bindegewebe. III. Die embryonale Histogenese des Knochenmarks der Säugetiere. Arch Mikrosk Anat 76:1−113

McCulloch EA, Till JE (1963) Repression of colony-forming ability of C 57 BL haematopoietic cell transplantates into nonisologous hosts. J Cell Comp Physiol 61:301−308

McCuskey RS, Meineke HA, Townsend SF (1972) Studies of the microenvironment. I. Changes in the microvascular system and stroma during erythropoietic regeneration and suppression in the spleens of CF 1 mice. Blood 39:697−712

McFadden KD (1966) A study of iron storage in the developing rat spleen through the use of the prussian blue reaction. Growth 30:325−332

McFadden KD (1967) Megakaryocytes in the rat spleen. Can J Zool 45:1035−1049

Messner HA (1967) Untersuchungen zur Proliferationskinetik der Erythropoese bei perniziöser Anämie mit [3] H-Thymidin. Inaug Diss, Ulm

Messner H, Fliedner TM, Cronkite EP (1969) Kinetics of erythropoietic cell proliferation in pernicious anemia. Ser Haematol II: 44−64

Metcalf D, Moore MAS (1971) Haemopoietic cells. North-Holland, Amsterdam London

Metcalf D, Stevens S (1972) Influence of age and antigenic stimulation on granulocyte and makrophage progenitor cells in the mouse spleen. Cell Tissue Kinet 5:433−446

Miller JFAP, Block M, Rowlands DT, Kind P (1965) Effect of thymectomy on hematopoietic organs of the opossum "embryo". Proc Soc Exp Biol Med 118:916−921

Miller SC, Osmond DG (1973) The proliferation of lymphoid cells in Guinea-pig bone marrow. Cell Tissue Kinet 6:259−269

Moffat DJ, Rosse C, Yoffey JM (1967) The identity of the haemopoietic stem cell. Lancet 2:547−548

Monette FC, LoBue J, Gordon AS, Alexander P (1968a) Erythropoiesis in the rat: differential rates of DNA synthesis and cell proliferation. Science 162:1132−1134

Monette FC, LoBue J, Chan PC, Gordon AS (1968b) DNA synthesis time and related parameters in erythroid cell precursors of rats. Scand J Haematol 5:325−332

Mori Y, Lennert K (1969) Electron microscopic atlas of lymphnode cytology and pathology. Springer, Berlin Heidelberg New York

Mosier DE, Johnson BM (1975) Ontogeny of mouse lymphocyte function II. Development of the ability to produce antibody is modulated by T lymphocytes. J Exp Med 141:216−226

Müller-Hermelink HK, Lennert K (1978) The cytologic, histologic and functional basis for a moder classification of lymphomas. In: Lennert K Malignant lymphomas other than Hodgkins disease. Handbuch der speziellen pathologischen und anatomischen Histologie, Springer, Berlin Heidelberg New York pp 1−70

Neumann E, Honetz N, Mittermayer K, Schwarzmeier JD (1976) Zellkinetische Untersuchungen bei Osteomyelosklerose. In: Stacher A, Höcker P (eds) Erkrankungen der Myelopoese, Leukämien, myeloproliferatives Syndrom, Polycythämie. Urban & Schwarzenberg, München Berlin Wien pp 37−39

Nicolescu P, Rouiller CH (1967) Beziehungen zwischen den Endothelzellen der Lebersinusoide und den von Kupfer'schen Sternzellen. Elektronenmikroskopische Untersuchungen. Z Zellforsch 76:313–338

North RJ (1971) Methyl green-pyronin for staining autoradiographs of hydroxyethyl metacrylate embedded lymphoid tissue. Stain Technol 46:59–62

Odartchenko N, Cottier H, Bond VP (1971) A study on ineffective erythropoiesis in the dog. Cell Tissue Kinet 4:107–112

Odell TT Jr, Burch EA Jr, Jackson CW, Friday TJ (1969) Megakaryocytopoiesis in mice. Cell Tissue Kinet 2:363–367

Orlic D, Gordon AS, Rohdin JAG (1965) An ultrastructural study of erythropoietin-induced red cell formation in mouse spleen. J Ultrastruct Res 13:516–542

Orlic D, Gordon AS, Rhodin JAG (1968) Ultrastructural and autoradiographic studies of erythropoietin-induced red cell production. Ann NY Acad Sci 149:198–216

Parott DMV, Sousa M de (1971) Thymus-dependent and thymus-independent populations: Origin, migratory patterns and lifespan. Clin Exp Immunol 8:663–684

Parry DM, Blackett NM (1972) Ineffective erythropoiesis in normal rats and "early labelled" bile pigment. Acta Haematol 47:348–355

Patt HM, Maloney MA (1964) A model of granulocyte kinetics. Ann NY Acad Sci 113:515–522

Picted R, Orci L, Forssmann WG, Girardier L (1969) An electron microscope study of the perfusion-fixed spleen. I. The spleen circulation and the RES concept. Z Zellforsch 96:372–399

Ploemacher RE, Soest van PL (1977a) Morphological investigation on phenylhydrazine-induced erythropoiesis in the adult mouse liver. Cell Tiss Res 178:435–461

Ploemacher RE, Soest van PL (1977b) Morphological investigation on ectopic erythropoiesis in experimental hemolytic anemia. Cytobiologie 15:391–409

Plum CM (1943) Zur Granulocytopoese bei Ratten. Folia Haematol 67:119–127

Porcellini A, Delfini C, Lucarelli G (1976) Kinetics of erythroid cell precursors in the newborn rat. Proc Soc Exp Biol Med 153:125–130

Prindull G (1966) Die Milz als nachgeordnetes lymphatisches Organ. Autoradiographische Studien zur Entwicklung der Milz bei der Maus. Z Anat Entwicklungsgesch 125:255–275

Prothero J, Starling M, Rosse C (1978) Cell Kinetics in the erythroid compartment of guinea pig bone marrow: a model based on ^{3}H-TdR studies. Cell Tissue Kinet 11:301–316

Queißer W (1978) Das thrombocytäre Zellsystem. In: Queißer W. (ed): Das Knochenmark. Morphologie – Funktion – Diagnostik. Thieme, Stuttgart pp 209–226

Ramsell TG, Yoffey JM (1961) The bone marrow of the adult rat. Acta Anat 47:55–65

Reincke U (1962) Über die Hämatopoese junger Wistar-Ratten. Arch Exp Veterinaermed 16:303–313

Resnitzky P, Zipori D, Trainin N (1971) Effect of neonatal thymectomy on hemopoietic tissue in mice. Blood 37:634–646

Renricca NJ, Rizzoli V, Howard D, Duffy P, Stohlmann F (1970) Cell emigration and proliferation during severe anemia. Blood 36:764–771

Renricca NJ, Howard D, Kubanek B, Stohlmann F Jr (1976) Erythroid differentiation of fetal, newborn and adult haemopoietic stem cells. Scand J Haematol 16:189–195

Richardson KC, Jaretta L, Finke EH (1960) Embedding in epoxy resins for ultrathin sectioning in electron microscopy. Stain Technol 35:313–323

Richter H (1953) Vergleichende Untersuchungen über das Hämo-, Myelo- und Splenogramm bei Tieren mit Stoffwechsel- und Speichermilz. Z Zellforsch 38:509–525

Riegel K, Ruhrmann G (1964) Über die Atemgastransportfunktion des Blutes und die Erythropoese junger Kaninchen. Acta Haematol 32:129–135

Rifkind RA, Chui D, Epler H (1969) An ultrastructural study of early morphogenetic events during the establishment of the fetal hepatic erythropoiesis. Cell Biol 40:343–365

Riman J, Seifert J, Vesely J (1958) Quantitative cytochemical studies concerning the growth of haemopoietic organs. Part I. Development of cellularity and enlargement of organ mass of the rat spleen during postnatal ontogenesis. Neoplasma 4:379–381

Rondanelli EG, Magliulio E, Giraldi A, Carco FP (1967) The chronology of the mitotic cycle of human granulocytopoietic cells. Phase contrast studies on living cells in vitro. Blood 30:557–565

Roser B (1970) The origins, kinetics and fate of macrophage populations. J Reticuloendothel Soc 8:139−161

Rosse C (1976) Small lymphocyte and transitional cell populations of the bone marrow; their role in the mediation of immune and hemopoietic progenitor cell functions. Int Rev Cytol 45: 155−290

Rosse C, Yoffey JM (1967) The morphology of the transitional lymphocyte in guinea-pig bone marrow. J Anat 102:113−124

Rosse C, Trotter JA (1974) Cytochemical and radioautographic identification of cells induced to synthesize hemoglobin. Blood 43:885−898

Roylance PJ (1968) An evaluation of erythropoiesis in the young rat. Cell Tissue Kinet 1:299−308

Ruhrmann G (1966) Morphologische Untersuchungen über die Erythropoese des Neugeborenen und des Säuglings. Z Ges Exp Med 140:319−367

Schultze B (1968) Die Orthologie und Pathologie des Nucleinsäure- und Eiweißstoffwechsels der Zelle im Autoradiogramm. In: Hdb. allg. Path. Bd. II/5, Stoffwechsel und Feinstruktur der Zelle. Springer, Berlin Heidelberg New York, pp 466−670

Silini G, Andreozzi U, Pozzi LV (1976) The role of the spleen in the repopulation of the haemopoietic system of heavily irradiated mice. Cell Tissue Kinet 9:341−350

Sjögren U (1978) Different composition of the erythropoietic tissue in bone marrow, spleen and liver in myelofibrosis. Acta Haematol 59:231−236

Sjögren U (1979) Morphologic studies of the erythropoietic part of the bone marrow in myeloid leukaemias. Scand J Haematol 22:61−66

Sjögren U, Brandt L (1976) Different composition and mitotic activity of the haemopoietic tissue in bone marrow, spleen and liver in chronic myeloid leukaemia. Acta Haematol 55:73−80

Söderström N, Bandmann U, Lundh B (1975) Patho-anatomical features of the spleen and liver. Clin Haematol 4:309−329

Sousa MAB de, Parrott DMW (1967) The definition of a germinal center area as distinct from the thymusdependent area in the lymphoid tissue of the mouse. In: Cottier H, Odartchenko N, Schindler R, Congdon CC (eds) Germinal centers in immune responses. Springer, Berlin Heidelberg New York, pp 361−370

Sprent J (1973) Circulating T and B lymphocytes of the mouse. I. Migratory properties. Cell Immunol 7:10−39

Starling MR, Rosse C (1976) Cell proliferation in the erythroid compartment of Guinea-pig bone marrow: studies with ^{3}H-thymidine. Cell Tissue Kinet 9:191−204

Stohlman F Jr, Lucarelli G, Howard D, Morse B, Leventhal B (1964) Regulation of erythropoiesis. XVI. Cytokinetic patterns in disorders of erythropoiesis. Medicine 43:651−660

Stutte HJ (1974) Hypersplenismus und Milzstruktur. Ferment-histochemische und biometrische Untersuchungen an menschlichen Milzen. Thieme, Stuttgart

Stutte HJ, Parwaresch MR, Rossius H (1974) Über die Regeneration von Milzautotransplantaten beim Kaninchen. Fermenthistochemische Untersuchungen. Res Exp Med 163:79−94

Tarbutt RG, Blackett NM (1968) Cell population kinetics of the recognizable erythroid cells in the rat. Cell Tissue Kinet 1:65−80

Tarbutt RG (1969) Cell population kinetics of the erythroid system in the rat. The response to protracted anaemia and to continuous γ-irradiation. Brit J Haematol 16:9−24

Tarbutt RG, Cole RJ (1970) Cell population kinetics of erythroid tissue in the liver of foetal mice. J Embryol Exp Morphol 24:429−446

Tavassoli M (1975) Studies on hemopoietic microenvironments. Exp Hematol 3:213−226

Theil EC (1980) Annotation. Embryonic erythropoiesis and iron metabolism. Brit J Haemat 45: 357−360

Till JE, McCulloch EA (1961) A direct measurement of the radiation sensitive of normal mouse bone marrow cells. Radiat Res 14:213−222

Tischendorf F (1969) Die Milz, Blutgefäß- und Lymphgefäßapparat, Innersekretorische Drüsen. In: Oksche A, Vollrath L (eds) Handbuch der mikroskopischen Anatomie des Menschen Bd VI/6 Springer, Berlin Heidelberg New York

Trainin N, Resnitzky P (1969) Influence of neonatal thymectomy on cloning capacity of bone marrow cells in mice. Nature 221:1154−1155

Trentin JJ (1971) Determination of bone marrow stem cell differentiation by stromal hemopoietic inductive microenvironments (HIM). Am J Pathol 65:621–628

Vacek A, Bartonickova A, Tkadlecek L (1976) Age dependence of the number of the stem cells in haemopoietic tissues of rats. Cell Tissue Kinet 9:1–8

Veermann AJP (1974) On the interdigitating cells in the thymus-dependent area of the rat spleen: a relation between the mononuclear phagocyte system and T-lymphocytes. Cell Tissue Res 148:247–257

Veermann AJP (1975) The postnatal development of the white pulp in the rat spleen and the onset of immunocompetence against a thymus-independent and a thymus-dependent antigen. Z Immunol Forsch 150:45–59

Warninghoff G, Hausmann K (1955) Die Morphologie der embryonalen Hämatopoese des Menschen im Vergleich zu postfetalen Blutbildungsstörungen. Acta Haematol 14:273–291

Warr GW, Sljivic VS (1974) Origin and division of liver macrophages during stimulation of the mononuclear phagocyte system. Cell Tissue Kinet 7:559–565

Weicker H (1954a) Exakte Kriterien des Knochenmarks: Die Maß- und Mengenrelationen der Erythroblasten als Ausdruck der Reifungs- und Teilungsgesetze der Erythropoese. Schweiz Med Wschr 8:245–251

Weicker H (1954b) Die Morphogenese der Anämien als Ergebnis der metrisch-kombinatorischen Analyse der Erythroblasten. Aerztl Wschr 9:1017–1024

Weicker H (1956) Ein quantitatives Modell der Granulopoese. Schweiz Med Wschr 86:1456–1460

Weicker H (1957) Die hemi-homoplastische Teilung des Proerythroblasten – die Lösung des Stammzellproblems der Erythropoese. Folia Haematol 74:49–64

Wekel HP, Adam W, Heimpel H (1970) Vergleichende nuclear-medizinische und histologische Untersuchungen über die extramedulläre Blutbildung in Milz und Leber bei Osteomyelosklerose. In: Lennert K, Harms D (eds) Die Milz, Struktur, Funktion, Pathologie, Klinik, Therapie Springer, Berlin Heidelberg New York, pp 93–99

Wickramasinghe SN, Chalmers DG, Cooper EH (1967) Disturbed proliferation of erythropoietic cells in pernicious anaemia. Nature 215:189–191

Wickramasinghe SN (1976) Effect of azathioprine on the cell cycle of haemopoietic cells. Scand J Haematol 16:38–40

Wiersbowsky A, Grouls V, Helpap B, Klingmüller G (1982 in press) An electron microscopic study of the development of the periarteriolar zone in the splenic white pulp of rats. Cell Tissue Res

Whitelaw DM, Batho HF (1972) The distribution of monocytes in the rat. Cell Tissue Kinet 5:215–225

Yamashita K, Helpap B (1974a) A staining method of epoxy-embedded lymphatic and hemopoietic tissues in autoradiogramms. Histochemistry. 40:275–279

Yamashita K, Helpap B (1974b) Zur Zellzusammensetzung der roten Milzpulpa von Mäusen. I. Autoradiographische Untersuchungen mit ^{3}H-Thymidin an Semidünnschnitten. Virchows Archiv [Cell Path] 16:331–346

Yamashita K, Helpap B (1974c) Zur Zellzusammensetzung der roten Milzpulpa von Mäusen. II. Elektronenmikroskopische und autoradiographische Untersuchungen an Megakaryozyten. Virchows Archiv [Cell Path] 16:347–361

Yamashita K, Helpap B (1974d) unpublished data

Zamboni L Electron microscopic studies of blood embryogenesis in humans. II. The hemopoietic activity in the fetal liver. J Ultrastruct Res 12:525–541

Subject Index

*Other Reviews of
Interest in This Series*

Volume 65
E. Pannese
**The Satellite Cells of the
Sensory Ganglia**
1981. 30 figures. IX, 111 pages
ISBN 3-540-10219-1

Volume 66
H.-M. Schmidt
**Die Artikulationsflächen der
menschlichen Sprunggelenke**
1981. 45 Abbildungen, 2 Tabellen.
VII, 81 Seiten
ISBN 3-540-10306-6

Volume 67
H. Wolburg
**Axonal Transport,
Degeneration and Regenera-
tion in the Visual System of
the Goldfish**
1981. 28 figures. IX, 94 pages
ISBN 3-540-10336-8

Volume 70
W. Pfaller
**Structure Function
Correlation on Rat Kidney**
Quantitative Correlation of Structure and
Function in the Normal and Injured Rat
Kidney
1982. 23 figures. VIII, 106 pages
ISBN 3-540-11074-7

Volume 71
L. Thuneberg
**Interstitial Cells of Cajal:
Intestinal Pacemaker Cells?**
1982. 94 figures. Approx. 120 pages
ISBN 3-540-11261-8

Volume 72
H. Breuker
**Seasonal Spermatogenesis
in the Mute Swan
(Cygnus olor)**
1982. 30 figures. Approx. 120 pages
ISBN 3-540-11326-6

Volume 73
G. Zweers
**The Feeding System of
the Pigeon
(Columba livia L.)**
1982. 45 figures. Approx. 90 pages
ISBN 3-540-11332-0

Volume 74
J. Altman, S. A. Bayer
**Development of the
Cranial Nerve Ganglia and
Related Nuclei in the Rat**
1982. 64 figures. Approx. 120 pages
ISBN 3-540-11337-1

Springer-Verlag
Berlin
Heidelberg
New York